Applied
Aerodynamics

Applied Aerodynamics

Editor

Nicolaos Sabella

Applied Aerodynamics

Edited by **Nicolaos Sabella**

ISBN: 978-1-68117-234-7
Library of Congress Control Number: 2016934766

© 2017 by
SCITUS Academics LLC,
www.scitusacademics.com
Box No. 4766, 616 Corporate Way,
Suite 2, Valley Cottage,
NY 10989

Preface

Aerodynamics is a branch of fluid dynamics concerned with studying the motion of air, particularly when it interacts with a solid object, such as an airplane wing. Aerodynamics is a sub-field of fluid dynamics and gas dynamics, and many aspects of aerodynamics theory are common to these fields. The term aerodynamics is often used synonymously with gas dynamics, with the difference being that "gas dynamics" applies to the study of the motion of all gases, not limited to air. Studying the motion of air around an object allows us to measure the forces of lift, which allows an aircraft to overcome gravity, and drag, which is the resistance an aircraft "feels" as it moves through the air. Everything moving through the air (including airplanes, rockets, and birds) is affected by aerodynamics.The rules of aerodynamics explain how an airplane is able to fly. Anything that moves through air reacts to aerodynamics. A rocket blasting off the launch pad and a kite in the sky react to aerodynamics. Aerodynamics even acts on cars, since air flows around cars. Understanding the motion of air around an object (often called a flow field) enables the calculation of forces and moments acting on the object. In many aerodynamics problems, the forces of interest are the fundamental forces of flight: lift, drag, thrust, and weight. Of these, lift and drag are aerodynamic forces, i.e. forces due to air flow over a solid body. Calculation of these quantities is often founded upon the assumption that the flow field behaves as a continuum. Continuum flow fields are characterized by properties such as flow velocity, pressure, density and temperature, which may be functions of spatial position and time. These properties may be directly or indirectly measured in aerodynamics experiments, or calculated from equations for the conservation of mass, momentum, and energy in air flows. Density, flow velocity, and an additional property, viscosity, are used to classify flow fields. This book entitled Applied Aerodynamics covers the numerous cases of stationary and non-stationary aerodynamics, which is of invaluable tool for academicians, researchers and professionals.

Table of Contents

CHAPTER 1

Aerodynamics of the Cupped Wings during Peregrine Falcon's Diving Flight

Benjamin Ponitz, Michael Triep, Christoph Brücker

Institute of Mechanics and Fluid Dynamics, TU Bergakademie Freiberg, Freiberg, Germany

ABSTRACT

During a dive peregrine falcons can reach velocities of more than 320 km/h and makes themselves the fastest animals in the world. The aerodynamic mechanisms involved are not fully understood yet and the search for a conclusive answer to this fact motivates the three-dimensional (3-D) flow study. Especially the cupped wing configuration which is a unique feature of the wing shape in falcon peregrine dive is our focus herein. In particular, the flow in the gap between the main body and the cupped wing is studied to understand how this flow interacts with the body and to what extend it affects the integral forces of lift and drag. Characteristic shapes of the wings while diving are studied with regard to their aerodynamics using computational fluid dynamics (CFD). The results of the numerical simulations via ICEM CFD and OpenFOAM show predominant flow structures around the body surface and in the wake of the falcon model such as a pair of body vortices and tip vortices. The drag for the cupped wing profile is reduced in relation to the configuration of opened wings (without cupped-like profile) while lift is increased. The purpose of this study is primarily the basic research of the aerodynamic mechanisms during the falcon's diving flight. The results could be important for maintaining good maneuverability at high speeds in the aviation sector.

KEYWORDS

Peregrine Falcon, Aerodynamics, Cupped Wings, CFD

1. INTRODUCTION

The peregrine falcon (Falco peregrinus) is one of the world's fastest birds. During horizontal flight, it reaches velocities of up to 150 km/h ([1] [2]) and even more than 320 km/h when nose-diving to attack its bird prey (e.g. [3] - [10]). Nearly all bird species can alter the shape of their wings and thus they can change their aerodynamic properties [11] [12] , a concept known as "morphing wing" [13] . During a dive, peregrines also alter the shape of their wings; while accelerating, they move them closer and closer to their body [10] . Several body shapes can be described as a classical diamond shape of the wings followed by a tight vertical tuck with a cupped-like profile of the frontal wing parts [10] [14] - [16] . Only at top velocities (up to at least 320 km/h) peregrines build a wrap dive vacuum pack, i.e. the wings are completely folded against the elongated body [17]. Peregrines are not only extremely fast flyers but also maintain remarkable maneuverability at high speeds. For instance, during courtship behavior they often change their flight path at the end of a dive, i.e. they turn from a vertical dive into a steep climb. This suggests that peregrines are exposed to high mechanical loads.

Although the nose-diving flight of peregrines has been investigated for numerous times, exact numerical flow simulations have not been carried out. Therefore, we investigated different wing configurations (opened wings and cupped wings extension [9]) of the peregrine falcon during diving flight. Both geometries are gained from a previous study by Ponitz et al. (2014) [18] at maximum diving velocities from a dam wall dive. The present study investigates the influence of the geometry change of the extended cupped wings configuration via numerical simulations. The results show that the cupped wings reduce both form and induced drag, while increasing the lift coefficient. This shows how fine the bird can tune the body forces by morphing the wing shape in diving flight conditions.

2. MATERIALS AND METHODS

2.1. Geometries and Models of the Falcon

2.1.1. Dam Wall Diving Flights of Real Falcons

The numerical simulation in this study is based on the specific diving flight condition which is gained from a previous study by [18] . A peregrine falcon was trained to dive from a dam wall and the dive was captured with a stereo high-speed camera system to reconstruct the 3-D flight path. Furthermore, a camera equipped with 400 mm zoom lenses was used to gain detail studies of the falcon body geometry. The region of interest shows three flight phases: (1) acceleration/diving phase, (2) transient phase with roughly constant speed and (3) deceleration phase. A maximum dive speed of 22.5 m/s was derived from the flight trajectory at flight phase (2). During the dive two significant body geometries are determined around the transient flight phase (2): one configuration without cupped wings (hereinafter called "opened wings") and one configuration with cupped wings [9] (see Figure 1). For the investigated flight situation an angle of attack $\alpha = 5°$ was determined for the equilibrium condition at the maximum diving speed of 22.5 m/s. For this flight situation wings remain stationary and do not flap.

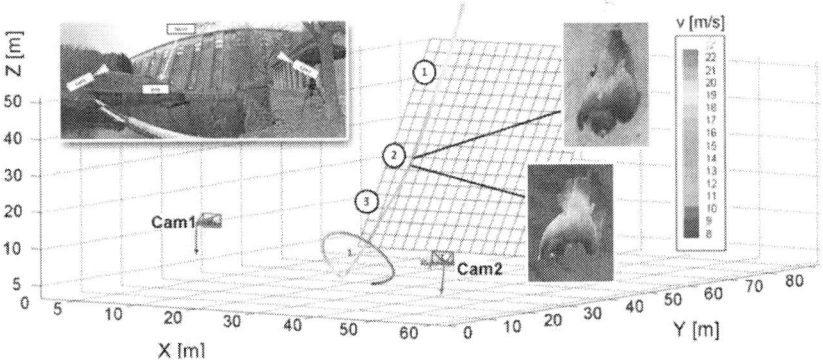

Figure 1. 3-D flight path and detailed body geometries of the falcon's dive from dam wall Olef-Talsperre, Germany. The trajectory is color-coded with the flight velocity magnitude (red color: higher velocities). The region of interest shows three flight phases: (1) acceleration/diving phase, (2) transient phase with roughly constant speed and (3) deceleration phase. Maximum velocity during the dive was 22.5 m/s at flight phase (2). For this transient phase two specific configurations (opened wings and cupped wings) were obtained. An angle of attack α = 5° was determined for the equilibrium condition [18] .

2.1.2. Generation of the Models

A life-sized model was built for both configurations, opened wings and cupped wings. Thus, we used the stuffed body of a female peregrine falcon and manually modified its wings until every body shape matched to the geometries of the falcon during dive from the dam wall (see Figure 1). The main part of the body for both configurations is identical. Only the tips of the wings are modified by a kind of winglets extension (cupped wings). Each modified body was fixated and subsequently scanned to acquire its 3-D surface contours (see Figure 2). In a further step, a one-to-one polyvinyl chloride (PVC) model was fabricated by laser sintering process using the acquired 3-D data. The experimental data from wind-tunnel test on this PVC falcon model is needed to evaluate the numerical simulations. Details of the aerodynamic relevant surface areas are given in Table 1.

2.2. Experimental Set-Up

For evaluating the numerical model the flow around the life-size PVC model was measured in a type Göttingenwind-tunnel. A force-balance delivered experimental data of the lift and drag forces of the identical model geometry. Hence, results of the wind-tunnel tests can be compared with numerical simulation results of lift and drag coefficients. The experimental set-up is depicted in Figure 3.

2.3. Numerical Set-Up

The numerical flow simulations of the falcon allow visualizing flow regions that are difficult to access by experimental methods. Furthermore, they enable to show distributions, for example, of the wall shear stress or pressure in order to identify hot spots, and most important for the present work, the exact determination of quantities like the λ_2 vortex criterion of the flow field. In the present study, the three-dimensional CAD model of the falcon is transferred into a computational unstructured grid using a grid generation tool ICEM CFD 14.5 (ANSYS, Inc., Canonsburg, PA, USA). The computational domain as marked in Figure 4 includes the inflow region, the falcon region, and the downstream wake region of the flow.

Special attention was paid to the meshing of the falcon. Refinements toward near-wall regions were taken into consideration. The grid consists in total of 6.5 million unstructured tetrahedron cells and 1.5 million prism cells on the falcon surface. A mesh independency check for the results of lift and drag coefficients was done for up to 10 million cells. Simulation stability was investigated in respect to different grid parameters and following settings leads to stable results: The height of the first prisms layer on the falcon surface is set to 0.1 mm $(y^+ = 0.2)$ with a growth factor of 1.10 for the following layer perpendicular to the wall and a total number of 10 layers (Figure 5). For these simulation parameters the results deliver stable values which furthermore match the experimental results of lift and drag forces obtained from the wind-tunnel tests (e.g. $C_{D,EXP} = 0.0698$ and $C_{D,CFD} = 0.0725$ for the drag coefficients of cupped wing model geometry). For determining the lift and drag coefficients we used the top-view projection area (seeTable 1) as the reference of the falcon model. Thus, the evaluated numerical simulation model is used for the following investigations.

The numerical flow simulation was performed using the open source CFD software OpenFOAM (OpenCFD Ltd., Bracknell, UK). The code numerically solves the conservation equations of mass and momentum by means of a finite volume approach. The Reynolds number Re based on the length $L_{ref} = 400$ mm reaches values of about $Re = 587000$ for $u = 22.5$ m/s. Therefore, turbulent flow is taken into account by a Reynolds averaged approach and the one equation Spalart-Almaras turbulence model. The turbulent viscosity v_T for the Spalart-Almaras model can be determined by:

$$\tilde{v}_T = \sqrt{\frac{3}{2}}u \cdot I \cdot l$$

$$(1)$$

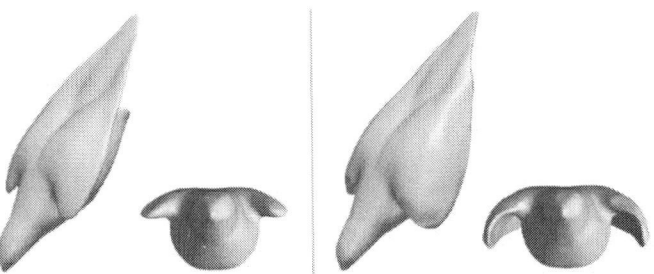

Figure 2. Computer aided design (CAD) falcon models: opened wingshape (left) and cupped wing shape (right).

Table 1. Reference areas of the falcon modelswith opened wings and cupped wings.

Aspect view	Reference area	
	Opened wings	Cupped wings
Top-view projection area	$A_{ref.opened.top} = 0.0411$ m^2	$A_{ref.cupped.top} = 0.0421$ m^2
Frontal projection area	$A_{ref.opened.front} = 0.0123$ m^2	$A_{ref.cupped.front} = 0.0139$ m^2

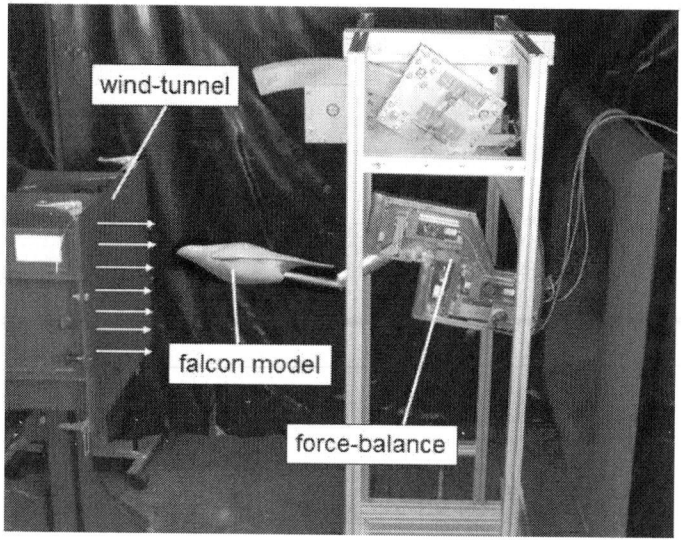

Figure 3. PVC falcon model and force-balance in the wind- tunnel [18] .

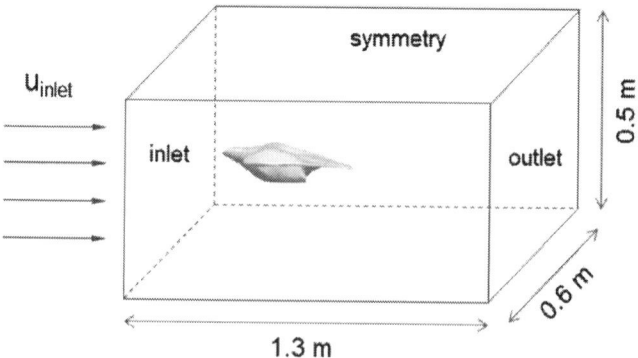

Figure 4. Numerical model dimensions and boundary condi- tions.

where u is the free stream velocity, I is the turbulence intensity ($I = 0.04\%$, from experiments [18]) and l is the length scale$(l = 0.07 L_{ref})$. The comparison with the results gained from the two equations k - ω SST model showed that the Spalart-Almaras simulations delivered more stable results. A major advantage of this model is that it was developed for flow simulations around an airfoil including wake regions and stall phenomena. Air was treated as a single-phase, incompressible$(Ma = 0.07)$, isothermal (20°C) Newtonian fluid with constant density (1.189 kg/m³) and viscosity$(18.232 \times 10^{-6}~\text{Pas})$. Boundary conditions were chosen in agreement with the experimental situation described in the section before and are defined in Table 2.

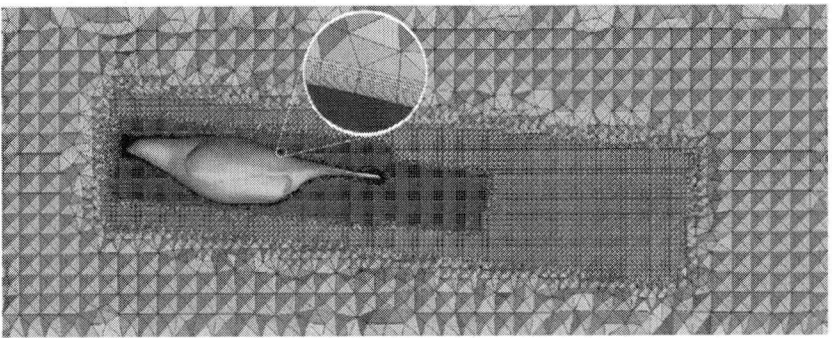

Figure 5. Prisms layer and density boxes of the grid refinement.

Table 2. Overview of boundary conditions.

Boundary	Flow variables		
	u	p	$\tilde{v}_r$
Falcon	$u = 0$	$n \cdot \nabla p = 0$	$\tilde{v}_r = 0$
Inlet	22.5 m/s	$n \cdot \nabla p = 0$	$3.08 \times 10^{-4}~\text{m}^2/\text{s}$
Outlet	$n \cdot \nabla u = 0$	1013×10^3 Pa	$n \cdot \nabla \tilde{v}_r = 0$
Sides	Symmetry	Symmetry	Symmetry

Due to the incompressible character of the fluid, the pressure was set in average constant in the outlet of the test compartment, so that the simulated relative pressure field can be transferred in the post-processing to the correct pressure level with the help of experimental measurements. The simulations were carried out for steady flow conditions.

3. RESULTS

The angle of attack for the investigated flight situation was determined in a previous study by Ponitz et al. (2014) [18] for the equilibrium condition at the maximum diving speed and is set to $\alpha = 5°$. Figure 6 shows a schematic vector diagram of the acting lift and drag forces on the falcon as well as the angle of attack in relation to the reference line of the falcon model which is built between the falcons tip and tail. The tip of the falcon is defined as the origin of the coordinate system (e.g. $x = 0$ m).

Results from the numerical simulation are post-processed with the software packages ParaView 4.0.1 (Kitware Inc., Clifton Park, NY, USA) and Tecplot 360 2013 (Tecplot Inc., Bellevue, WA, USA). Thus, the integral forces such as lift and drag forces are calculated from the velocity and pressure data field and visualizations are realized. With the calculated lift (L) and drag (D) forces the associated coefficients could be determined as followed:

$$C_L = \frac{L}{q \cdot A} = \frac{2 \cdot L}{\rho \cdot u^2 \cdot A}$$

(2)

And

$$C_D = \frac{D}{q \cdot A} = \frac{2 \cdot D}{\rho \cdot u^2 \cdot A},$$

(3)

where q is the dynamic pressure, ρ is the mass density of the fluid, u is the free stream velocity and A is the reference area of the wing, in both cases the top-view projection area $A_{ref,top}$.

All results are illustrated as comparative data between both configurations, the opened wing shape and the cupped wing shape. Table 3 shows the obtained lift and drag coefficients. Figure 7 shows the visualizations between the two configurations for surface pressure distribution, wall shear stress, streamlines below the wings, wake flow via λ_2 vortex detection criterion [19] and slices in the wake [20] . In addition, Figure 8 gives a more detailed insight of the surface streamlines on both body configurations.

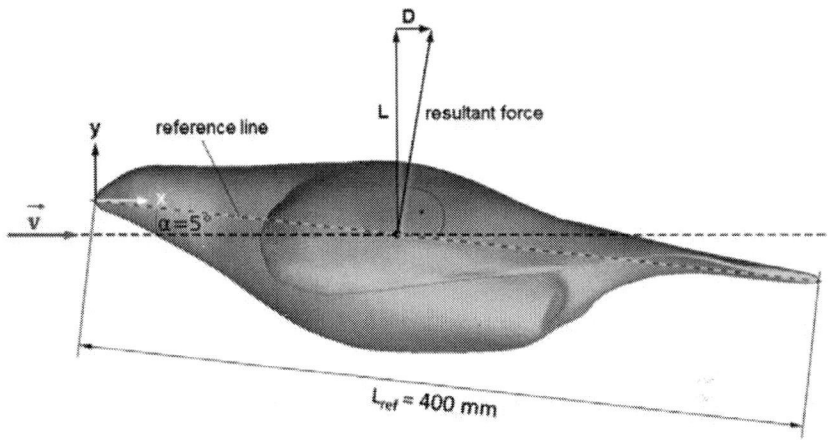

Figure 6. Definition of reference line, angle of attack and schematic vector diagram of forces.

Table 3. Comparison of lift and drag coefficients for both body geometries: opened wings vs. cupped wings (Reference area is the top-view projection area).

Flow parameter	Falcon model configuration	
	Opened wings	Cupped wings
Lift coefficient C_L	0.0851	0.1119
Drag coefficent C_D	0.0892	0.0725

The comparison of the lift coefficients (Table 3) shows a value of $C_{L,opened} = 0.0851$ for the configuration with opened wings and a value of $C_{L,cupped} = 0.1119$ for the cupped wing shape. The cupped wing profile causes an increased lift coefficient for identical simulation parameters. This is due to the cupped wings which deliver more wing area for lift. The passing air flow is held below the wing on the pressure side of the cupped wing. Hence, the acting lift component is increased in relation to the geometry of opened wing configuration. More interesting is the focus on the drag coefficient. Due to the extended cupped wing shape an increased drag coefficient is expected in relation to the opened wing geometry, however the results show the contrarian. With the cupped wings extension the drag coefficient is reduced from $C_{D,opened} = 0.0892$ down to a value of $C_{D.cupped} = 0.0725$, see Table 3. The reason could be found in the flow details as discussed below by means of streamlines and near-surface flow visualizations.

For comparison with bluff body aerodynamics we calculated in addition the drag coefficient with the projection area along the axis of flight (frontal projection area $A_{ref,front}$).). The following values were determined: $C_{D.opened,frontal} = 0.2980$ and $C_{D.cupped,frontal} = 0.2402$. They show the same tendency of decreased drag in the cupped wing configuration.

In the upper part of Figure 7 surface pressure distributions show typical areas of higher pressure on the tip of the falcon body and on the leading edge of the wings. Typical areas of lower pressure are found on the suction side of both wing configuration and especially between the cupped wings and the falcon body. In addition, wall shear stress distributions combined with the Line-Integral-Convolution (LIC) visualization method [21] show characteristic patterns of a streamlined body. Besides the visualizations of surface pressure and wall shear stress distributions, the streamlines and wake flow characteristics below and behind the falcon models show significant flow features (see middle and lower part of Figure 7). For instance, the comparison of the streamlines below the wings (color coded with the velocity magnitude) show different pathways and various hotspots of maximum velocity values below the wings. Streamlines around the opened wing configuration indicates flow separation whereas the cupped wings show streamlines which lies close to the body contour for a much longer distance downstream. Additionally, the spot of higher velocity magnitude below the cupped wings indicate an acceleration of the flow in this region like in a tapering nozzle. Therefore the flow remains attached to the body for a longer distance downstream, which affects the body drag in a beneficial way. This could be one reason for the reduced drag of the cupped wing geometry in relation to the opened wing shape.

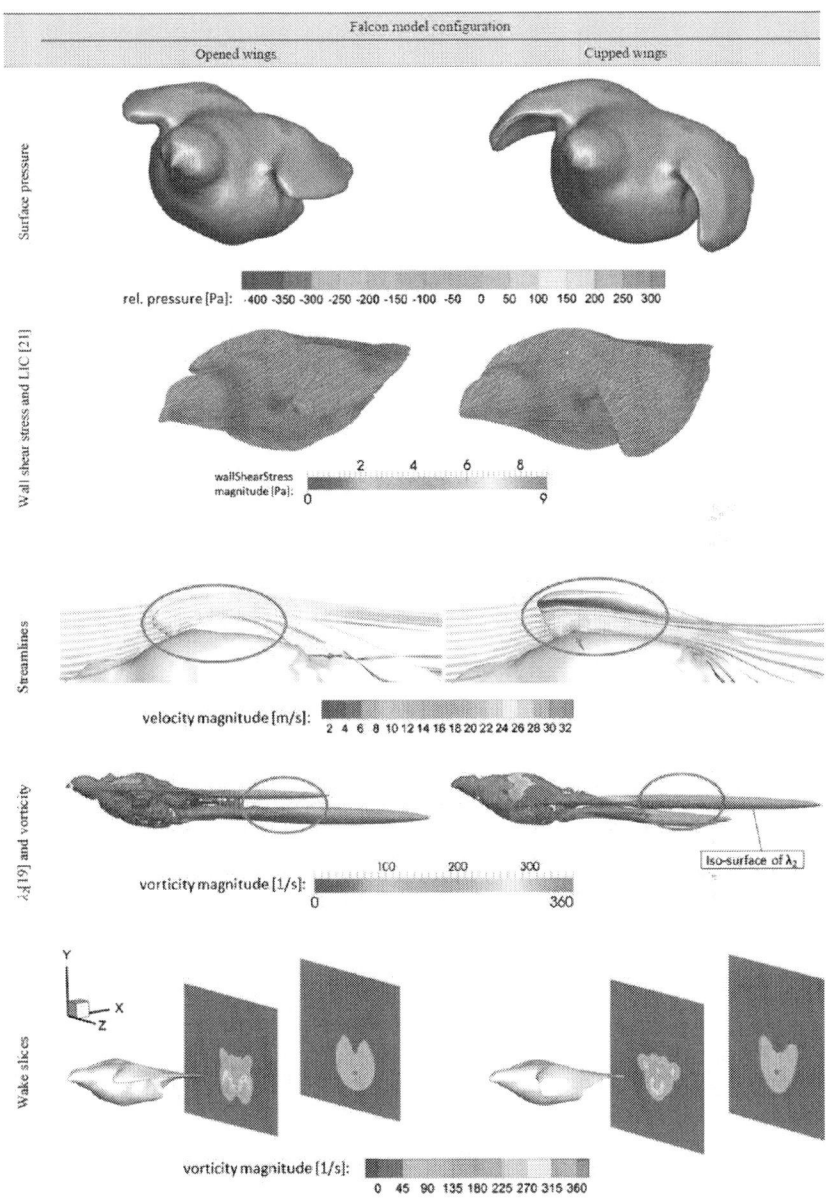

Figure 7. Comparison of visualized flow parameters for both configurations: opened wings vs. cupped wings.

Furthermore, the wake flow behind the model visualized by the λ_2 vortex detection criterion (iso-surface is color coded with the vorticity magnitude) let recognize some differences between both configurations. In both cases two vortex pairs are recognizable. One vortex pair is generated around the wings (known as the wing-tip vortex) and one vortex pair is generated from flow separation at the aft part of the falcon main body (called herein the body vortex). The vorticity distribution in the wake is shown for two discrete slices in the Trefftz-   plane   [20]   at   the   positions   close to $(x=0.5$ m$)$ and far behind $(x=0.85$ m$)$ the falcon model. The cupped wings lead to wing-tip vortices which are located further down in vertical direction than in the case of the opened wing configuration. Body vortices of both model shapes occur rather in the same location. Hence, the spatial arrangement of the vortex pairs (wing-tip and body vortices) is significantly closer in the case of the cupped wings geometry. In general, the induced drag of a wing depends on the strength of circulation and the lateral position of the wake vortices away from the centerline, thus it is concluded that the observed differences also influence the induced drag for both geometries.

Figure 8 depicts details of the surface streamlines around both body configurations. Streamlines below the opened wings indicates more clearly local flow separation in relation to the geometry with cupped wings where streamlines appear more continuous such as in an attached flow.

Figure 9 illustrates the three-dimensional arrangement of wake vortices visualized by the iso-surface of the vortex detection criterion λ_2 colored coded with the vorticity magnitude. The velocity distribution in the slice at position $x = 0.64$ m shows the downwash in the wake region of the falcon.

4. DISCUSSION AND CONCLUSIONS

This study investigated in detail the aerodynamics of steady flight conditions of a peregrine falcon in dive motion. The contours of body and wing shape as well as the dive speed and angle of attack have been detected in a previous study and were used herein for numerical flow simulations around the body and in the wake: we simulated the flow at an angle of attack of 5° and a flow speed of 22.5 m/s.

The focus of this study was the comparison of the opened wing shape and geometry with cupped extension of the wing tip. The formulation "cupped wings" was first used by [9] to describe the shape of the downward tilted tips of the wing which is a typical falcon shape during dive. When comparing both wing configurations the results clearly reveal that the cupped wing configuration increases lift and decreases drag under the same flow conditions (and angle of attack). Although the total surface area in the cupped wing configuration is larger, the body drag is reduced. The reason therefore is the acceleration of the flow close to the body in the gap between cupped wing tip and main body surface. Therefore flow separation is shifted towards the trailing edge of the body which reduces the form drag of the body. In addition, the modification of the tip vortices position and strength in the wake of the cupped wings hints on the beneficial effect on reducing the induced drag. This can be deduced from the analysis of the flow in the Trefftz-plane behind the body. This shows how fine the bird can tune the body forces by morphing the wing shape in diving flight conditions.

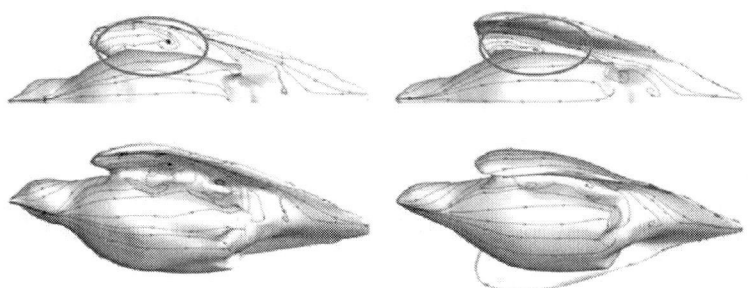

Figure 8. Comparison of visualized surface streamlines for both configurations: opened wings (left) vs. cupped wings (right).

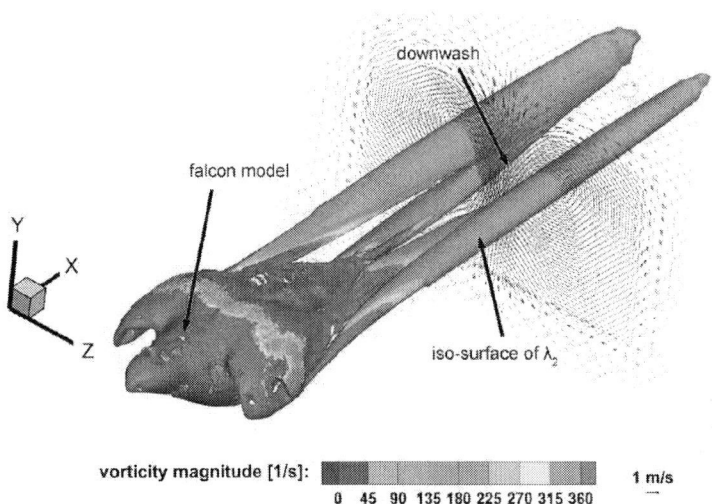

Figure 9. Three-dimensional visualization of the spatial arrangement of wake vortices. The iso-surface of λ_2 criterion is color coded with the vorticity magnitude.

Conclusions drawn herein are based on a smooth surface of the model. In nature the body is covered with feathers which may also play a role [22] . The tiny scales and the elastic properties of the feathers were not taken into account. This is subject of ongoing work.

ACKNOWLEDGEMENTS

The study is funded by the Deutsche Forschungsgemeinschaft (BL 242/19-1; BR 1494/21-1). The funders had no role in study design, data collection and analysis, decision to publish, or preparation of the manuscript. We thank Horst Bleckmann for his support in gathering the data of the cupped wing model.

REFERENCES

1. del Hoyo, J., Elliott, A., Sargatal, J. and Collar, N.J. (1999) Handbook of the Birds of the World. Vol. 5, Lynx Edicions, Barcelona.

2. Podbregar, N. (2013) Das Geheimnis des Fliegens—Tierischen Flugkünstlern auf der Spur: Strategien der Evolution. Springer, Berlin and Heidelberg, 227-243.

3. Tucker, V.A. and Parrott, G.C. (1970) Aerodynamics of Gliding Flight in a Falcon and Other Birds. Journal of Experimental Biology, 52, 345-367.

4. Orton, D.A. (1975) The Speed of a Peregrine's Dive. The Field, 588-590.

5. Brown, L.A. (1976) British Birds of Prey. Collins, London.

6. Alerstam, T. (1987) Radar Observations of the Stoop of the Peregrine Falcon Falco Peregrinus and the Goshawk Accipiter Gentilis. Ibis, 129, 267-273.

7. Savage, C. (1992) Peregrine Falcons. Sierra Club, San Francisco.

8. Clark, W.S. (1995) How Fast Is the Fastest Bird? WildBird, 9, 42-43.

9. Tucker, V.A. (1998) Gliding Flight: Speed and Acceleration of Ideal Falcons during Diving and Pull Out. Journal of Experimental Biology, 201, 403-414.

10. Franklin, D.C. (1999) Evidence of Disarray amongst Granivorous Bird Assemblages in the Savannas of Northern Australia, a Region of Sparse Human Settlement. Biological Conservation, 90, 53-68.

11. Nachtigall, W. (1975) Vogelflügel und Gleitflug Einführung in die aerodynamische Betrachtungsweise des Flügels. Journal für Ornithologie, 116, 1-38.

12. Nachtigall, W. (1998) Der Gleitflug von Vögeln. Physik in unserer Zeit, 1, 25-29.

13. Lentink, D., Müller, U.K., Stamhuis, E.J., de Kat, R., van Gestel, W., Veldhuis, L.L.M., et al. (2007) How Swifts Control Their Glide Performance with Morphing Wings. Nature, 446, 1082-1085.

14. Ratcliffe, D.A. (1980) The Peregrine Falcon. Buteo Books, Vermillion.

15. Hustler, K. (1983) Breeding Biology of the Peregrine Falcon in Zimbabwe. Ostrich, 54, 161-171.

16. Tucker, V.A. (1990) Body Drag, Feathers Drag and Interference Drag of the Mounting Strut in a Peregrine Falcon, Falco peregrinus. Journal of Experimental Biology, 149, 449-468.

17. Seitz, K. (1999) Vertical Flight. NAFA Journal, 38, 68-72.

18. Ponitz, B., Schmitz, A., Fischer, D., Bleckmann, H. and Brücker, C. (2014) Diving-Flight Aerodynamics of a Peregrine Falcon (Falco peregrinus). PLoS ONE, 9, e86506.

19. Jeong, J. and Hussain, F. (1995) On the Identification of a Vortex. Journal of Fluid Mechanics, 285, 69-94.

20. Krasny, R. (1987) Computation of Vortex Sheet Roll-Up in the Trefftz Plane. Journal of Fluid Mechanics, 184, 123- 155. http://dx.doi.org/ 10.1017/S0022112087002830

21. Cabral, B. and Leedom, L.C. (1993) Imaging Vector Fields Using Line Integral Convolution. In: Proceedings of ACM SIGGRAPH'93, Anaheim, 2-6 August 1993, 263-270.

22. Schmitz, A., Ponitz, B., Brücker, C., Schmitz, H., Herweg, J. and Bleckmann, H. (2014) Morphological Properties of the Last Primaries, the Tail Feathers, and the Alulae of Accipiter nisus, Columba livia, Falco peregrinus, and Falco tinnunculus. Journal of Morphology, Early View.

NOMENCLATURE

α	angle of attack [°]		q	dynamic pressure [Pa]
A_{ref}	reference area of the wing [m²]		ρ	mass density of the fluid [kg·m⁻³]
C_D	drag coefficient [-]		Re	Reynolds number [-]
C_L	lift coefficient [-]		u, v	free stream velocity [m·s⁻¹]
D	drag force [N]		v_r	kinematic viscosity [m²·s⁻¹]
I	turbulence intensity [%]		x	coordinate in flow direction [m]
L	lift force [N]		y^+	dimensionless wall distance [-]
λ_2	vortex detection criterion [s⁻¹]			
l	length scale for Spalart-Almaras [m]		CAD	computer aided design
L_{ref}	reference length of the model [m]		CFD	computational fluid dynamics
Ma	Mach number [-]		LIC	line integral convolution
p	pressure [Pa]		PVC	polyvinyl chloride

CHAPTER 2

Experimental Study on Influence of Pitch Motion on the Wake of a Floating Wind Turbine Model

Stanislav Rockel [1,*], Elizabeth Camp [2], Jonas Schmidt [3], Joachim Peinke [1], Raul Bayoán Cal [2] and Michael Hölling [1]

[1] *ForWind, University of Oldenburg, Ammerländer Heerstr. 136, 26129 Oldenburg, Germany*
[2] *Department of Mechanical and Materials Engineering, Portland State University, P.O. Box 751, Portland, OR 97207, USA*
[3] *Fraunhofer IWES, Ammerländer Heerstr. 136, 26129 Oldenburg, Germany*

ABSTRACT

Wind tunnel experiments were performed, where the development of the wake of a model wind turbine was measured using stereo Particle Image Velocimetry to observe the influence of platform pitch motion. The wakes of a classical bottom fixed turbine and a streamwise oscillating turbine are compared. Results indicate that platform pitch creates an upward shift in all components of the flow and their fluctuations. The vertical flow created by the pitch motion as well as the reduced entrainment of kinetic energy from undisturbed flows above the turbine result in potentially higher loads and less available kinetic energy for a downwind turbine. Experimental results are compared with four wake models. The wake models employed are consistent with experimental results in describing the shapes and magnitudes of the streamwise velocity component of the wake for a fixed turbine. Inconsistencies between the model predictions and experimental results arise in the floating case particularly regarding the vertical displacement of the velocity components of the flow. Furthermore, it is found that the additional degrees of freedom of a floating wind turbine add to the complexity of the wake aerodynamics and improved wake models are needed, considering vertical flows and displacements due to pitch motion.

KEYWORDS

wake; wind tunnel; experiment; floating wind turbine; stereo particle image velocimetry (PIV); unsteady aerodynamics; wake models; offshore

1. INTRODUCTION

Wind energy has become a major contributor to energy from renewable sources and is still projected to increase its portion to the overall energy supply in the future. Offshore wind energy is found to have the highest potential to fulfill these demands, due to sustained winds which are unaffected by complex terrain [1]. Offshore wind turbines, which have been installed recently, use monopiles or tripods as foundations, which are feasible in shallow water up to a depth of 50 m. Such shallow areas are rare and are often already exploited [2], therefore, solutions have to be found to produce wind energy in deeper water areas. Floating support structures for offshore wind turbines are possible alternatives to current bottom fixed foundations. A limited amount of field and experimental data in regards to the operational conditions of floating wind turbines is available.

The additional degrees of freedom of a floating platform will cause different operational conditions compared to fixed foundations. A detailed understanding of the influence of the added degrees of freedom on aerodynamic performance, fatigue loads and finally the costs of floating wind turbines would allow an optimized design of floating wind turbines [3].

The influence of various motions such as pitch, sway and heave on the aerodynamics of the rotor and the wake characteristics are important in determining the design space for proper load and fatigue calculations [4]. Most wind turbines are operated in arrays so proper modeling of wake development plays an important role in order to reduce fatigue loads of turbines positioned downstream and maximize power output of the turbine ensemble [5]. Experimental data is needed to validate the performance of computational models.

Feasibility studies have been performed to define the constraints for floating turbines installations [6]. The critical constraints as outlined are to establish a design basis for offshore turbines, reliable offshore substructures as well as low cost anchors and moorings. Simulations of the structural response have been performed by Jonkman *et al.* [7,8]. The approach used consisted of expanding the FAST (Fatigue, Aerodynamics, Structures and Turbulence) code for classical turbines with hydrodynamic wave-body interaction programs such as WAMIT (Wave Analysis at Massachusetts Institute of Technology), that covers hydrodynamic damping and wave excitation of the platform. A comparison of three concepts of floating platforms with land-based turbines showed an overall increase of loadings on all turbine components due to the floating platform, so a mechanically more robust design is required [9]. Further combinations of platform and turbine design have to be tested and economic aspects must be considered in order determine an optimal design.

Matsukama *et al.* [10] performed motion analysis of a spar floating wind turbine under steady wind conditions. The effect of rotor rotation on the response of the floating platform was taken into account using blade element momentum theory and multibody dynamics theory, thus splitting the turbine into rotor, nacelle, tower and platform. It was found that motions like sway, roll and yaw are influenced by the gyroscopic moment of the rotor and that these motions drive variations in loads and power output of a turbine. A 9° static pitch of the platform was calculated for the rated wind speed of the turbine. The dynamic pitch of the platform was found to be the same on average, but oscillating around this value. Yaw and roll movement were estimated as well and found to drive the loads on the turbine and changes in variation in the power output.

Experimental investigations have been done on the structural response of floating support structures by Utsunomiya *et al.*[11]. The equation as posed by Morison, which accounts for inertia and drag force on a body in a flow, was used to determine the wave force on a the spar body and to estimate response amplitudes of a spar bouy. Qualitative agreement was found between estimates from the equation and experiments performed in a wave basin but the need for further improvements was stressed. A review of current literature on floating wind turbines is given in [12].

A preliminary study using the time averaged Unsteady Reynolds Averaged Navier-Stokes (URANS) method to simulate the aerodynamic interaction of the flow and pitch platform motion was performed by Matha *et al.* [13]. The effects of the mooring system dynamics on the turbine wake was explored for a case where the turbine pitched upwind and downwind in a uniform flow. Findings indicated that the wake of the floating turbine is susceptible to the floating conditions and the dynamics of the surrounding waves. Furthermore, results showed a stronger expansion of the wake and that vortices shed from the blade led to a rippled boundary layer in the upper wake.

Sebastian and Lackner [14] found current analysis methods based on blade element momentum theory did not capture the unsteady flow generated at the blades by strong variations in angle of attack. Large angle of attack changes were due to the additional motion of the floating platforms. These aerodynamic analysis methods use ad hoc formulations to cover the unsteadiness due to tip losses, high tip speed ratios and yawed flows. Motion induced unsteadiness violates assumptions of standard blade element momentum theory and leads to inaccurate predictions of unsteady aerodynamic loads. Spectral analysis of the kinematic behavior of various types of floating platforms were performed and the pitch aerodynamics was found to dominate unsteady flow effects and structural loading. It was shown that pitching motions of the wind turbine cause the turbine to change from the windmill state, where the turbine extracts energy from the flow, and the propeller state, under which the turbine drops energy into the flow [4]. The state of the turbine can be determined by observing the bound vorticity of the blades, where high vorticity represents a windmill state and low, negative vorticity represents a propeller state. It was found that vorticity change follows pitch motions of the platform and the variation in bound vorticity will lead to varying loads and fatigue on the rotor.

Jonkman et al. [7] performed simulations of fully coupled aero-hydro-servo-elastic responses of floating platform concepts and compared the results in special cases with other simulation frequency-domain analysis methods to test their reliability. None of the simulation results were validated against experimental data since no such data was freely available.

Wake measurements of offshore wind turbines were performed and compared with six different wake models [15]. The results showed a wide spread of predicted wake deficits.

The objective of this study is to investigate the influence of pitch motion on the wake of a model wind turbine. Wake measurements have been performed in a wind tunnel using stereoscopic particle image velocimetry (SPIV). The flow field, including mean and fluctuating components of the flow, are analyzed for a fixed versus an oscillating turbine under the same inflow. These results are then compared with existing wake models to determine their ability to capture such pitch motion effects.

2. THEORY AND MODELS

2.1. Mean Momentum and Kinetic Energy Equations

For steady, incompressible and inviscid flows, the Reynolds-averaged Navier-Stokes (RANS) equations are given by:

$$U_j \frac{\partial U_i}{\partial x_j} = -\frac{1}{\rho} \frac{\partial P}{\partial x_i} - \frac{\partial \overline{u_i u_j}}{\partial x_j} - F_{x_i} \qquad (1)$$

where x_i is the coordinate in the streamwise, wall-normal or spanwise direction, x, y, z, respectively; and U_i is the velocity vector, represented by its components U, V, W in the streamwise, wall-normal or spanwise (orthogonal to x and y) direction. The mean pressure is represented by P while u_i represents the particular fluctuating velocity component and ρ is the fluid density. The components of the thrust force of the wind turbine on the flow are represented by F_{xi}. The overbar denotes a time average. Unsteady and viscous term are omitted here, as the flow is considered to be steady and far from solid boundaries, where the viscosity plays an important role [16,17]. The momentum in Equation (1) yields a balance between the inertial terms, pressure gradient, Reynolds stress and the thrust force.

Multiplication of Equation (1) by the mean velocity U_i results in the mean kinetic energy equation:

$$U_j \frac{\partial \frac{1}{2} U_i^2}{\partial x_j} = -\frac{1}{\rho} U_i \frac{\partial P}{\partial x_i} + \overline{u_i u_j} \frac{\partial U_i}{\partial x_j} - \frac{\partial \overline{u_i u_j} U_i}{\partial x_j} - U_i F_{x_i} \qquad (2)$$

where the sum of $\frac{1}{2} U_i^2$ describes the mean kinetic energy in the flow. Equation (2) shows that the convection of mean kinetic energy equals a sum of the mean pressure gradient in the flow, the production of mean kinetic energy, the gradient of the kinetic energy flux and power extracted by the wind turbine, respectively.

Several terms of the momentum in Equation (1) and the mechanical energy in Equation (2) will be analyzed and related to the extraction of power, as it has been found that their balance yields the power production [16,17]. For further details on the derivation, see [16].

2.2. Wake Models

Both the mean momentum in Equation (1) and the mean kinetic energy in Equation (2) contain the Reynolds stress tensor $\overline{u_i u_j}^*$. As a consequence of the dynamic equation that can be derived for this quantity, the Reynolds stress tensor is not fully determined by the averaged velocity U_i and pressure P alone [18]. In addition, information about the average over products of three velocity fluctuation components $\overline{u_i u_j u_k}$ in the Reynolds stress equation is needed, otherwise called the triple correlation. The fact that this problem repeats for all moments of the velocity fluctuation is the closure problem of turbulence. This issue is circumvented by the application of a turbulence model which imposes a closed expression for $\overline{u_i u_j}$ in terms of U_i, P and possibly additional variables. These additional variables are often governed by partial differential equations similar to Equation (1) in that such equations balance convection with source, sink and stress terms.

Wake models that solve systems of partial differential equations are often summarised as field models. While the numerical solution of the continuity and momentum equations in three dimensions with a turbulence model and viscosity effects is obtained in computational fluid dynamics (CFD), also systems of reduced complexity are often considered. The model by Ainslie [19] is often employed, which imposes axial symmetry and an eddy viscosity model for turbulence closure. Consequently, this allows the formulation of a closed, two dimensional system of partial differential equations describing the far wake behind a single wind turbine.

Further simplified descriptions are given by analytical wake models. These follow from momentum balance considerations in a specific control volume, together with model-specific assumptions. One widely used example is the model by Jensen [20], which relies on the restriction to one velocity component and linear wake expansion. Often such models include parameters that are determined empirically. Note that if enough data is available, it is also possible to build wake models that are entirely empirical [21].

Herein, short summaries of the Jensen model, the Larsen model, the Ainslie model and a CFD actuator disk model are provided, before comparing the results to data obtained from wind tunnel experiments. The governing equations of the wake models can be found in Appendix A.

For the calculation of the wake models, *flapFOAM* [22] is used. This follows a similar approach as the wind farm layout program FLaP [23], but is fully embedded into the framework of OpenFOAM [24], thus extending the modeling capabilities. The resulting velocity deficits were added to the measured inflow profile, which was extended for low and high vertical coordinates by fitting standard log-profiles:

$$U(y) = \frac{u_\star}{\kappa} \ln \left(\frac{y + y_0}{y_0} \right) \tag{3}$$

where y denotes the wall-normal coordinate; $k = 0.41$ denotes the Karman constant; $u_\star$ denotes the friction velocity; and y_0 denotes the roughness length.

2.2.1. Jensen Model

The Jensen model wake is characterised by linear wake expansion as a function of the downstream distance from the rotor plane [20], with a proportionality constant usually chosen in the range of $k = 0.04$–0.07. Only the axial velocity deficit is modeled, with the magnitude obtained by momentum conservation. The deficit profile is hat-shaped in that it is constant in the radial direction within the wake and abruptly drops to zero at its boundary.

2.2.2. Larsen Model

The Larsen model [25] assumes an axisymmetric wake, reducing the RANS equations to two dimensions. For the Reynolds stresses, Prandtl's mixing length theory is used to model the eddy viscosity [26]. Model constants are determined empirically using measurements at $9.5D$ behind an isolated turbine. The wake radius is given as an explicit non-linear function of the downwind coordinate in this model. Descriptions of the horizontal and radial velocity components are given, which possess dependencies on the downstream and radial coordinates. The wake expansion and decay are determined by the thrust coefficient and the ambient turbulence intensity. The Larsen model has analytical solutions at first and second order with respect to an expansion in the axial velocity deficit.

2.2.3. Ainslie Model

Like the Larsen model, the Ainslie model [19] is based on the assumption of an axisymmetric wake. Viscous effects are taken into account using an eddy-viscosity model to obtain turbulence closure. Up to a distance of $2D$ downwind from the rotor the wake is prescribed using to a Gaussian profile, before it starts evolving according to the two coupled partial differential equations in the radial and downwind coordinates. Chosen parameters are the thrust coefficient and ambient turbulence, the latter entering the eddy viscosity components. The solutions for the axial and radial velocity components are obtained numerically on a two-dimensional grid.

2.2.4. Actuator Disk Rotor Model

Discretised versions of the RANS equations can be solved on a three-dimensional mesh with methods from computational fluid dynamics. The simplest representation of the rotor is the actuator disk approach with uniform forcing [27]. For turbulence modeling, *i.e.*, modeling the eddy viscosity to estimate Reynolds stresses, the $k - \varepsilon$ model with increased dissipation near the rotor is applied [28]. The inflow boundary condition for the velocity is described in Section 2.2. Likewise, the measured inflow profile for the turbulent kinetic energy (*TKE*) is extrapolated linearly for low and high vertical coordinates. The inflow turbulent dissipation ε could not be measured with sufficient accuracy, therefore it is interpolated using the Richards-Hoxey solution [29], which corresponds to the local log-profile approximation. In the simulations, all effects due to laboratory walls, apart from the ground, were ignored. The latter is represented by wall functions with roughness length $z_0 = 0.0046$ m.

3. METHODS

3.1. Wind Tunnel Facility

The experiments were performed in the wind tunnel at Portland State University. The test section of this closed-circuit tunnel is 5 m in length with a height of 0.8 m and a width of 1.2 m. The free stream velocity can be set between 2 m/s and 40 m/s with low turbulence intensity. Figure 1 shows a schematic view of the wind tunnel and setup used during the experiments. The inflow conditions were modified by a passive grid to generate a mean turbulence intensity level of 9%. Behind the grid, vertical strakes made of plexiglas were placed, shaped in such way, that the velocity profile resembles atmospheric-like conditions. For more details on the wind tunnel see [30].

The hub height of the turbine was 25 cm. The wind tunnel speed was set to 6.05 m/s at hub height of the turbine and the model wind turbine was placed 2.01 m downstream of the passive grid.

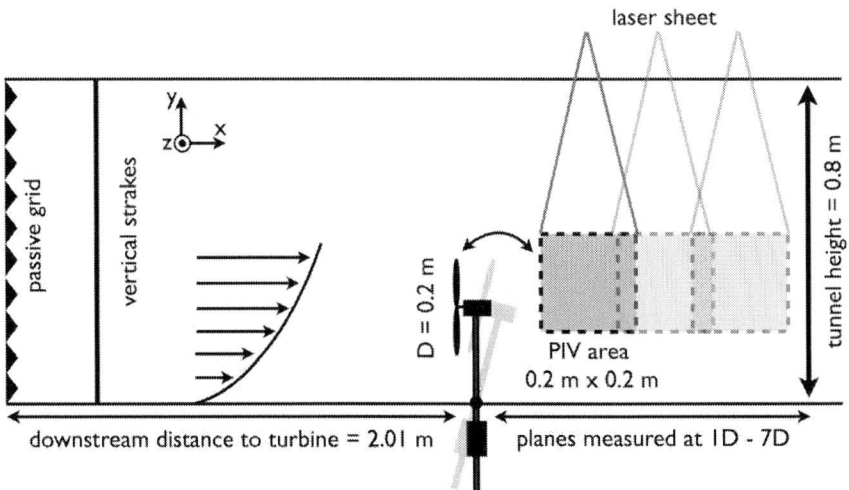

Figure 1. Wind tunnel setup. The scales are changed for visual clarification.

3.2. Model Wind Turbine

The model wind turbine shown in Figure 2 is a 1:400 scale model of a typical horizontal axis wind turbine with a 80 m rotor diameter. The model turbine consists of an aluminum tower with a diameter of 16 mm and a nacelle with a diameter of 28 mm. The rotor has a diameter D of 200 mm and is produced with a rapid prototyping method. The blockage ratio of rotor area to wind tunnel cross-section is below 3.3% so blockage effects can be neglected [31]. The blades are designed using Blade Element Momentum theory to perform efficiently at low wind speeds and therefore at low Reynolds numbers. The design tip speed ratio λ is approximately 6 which corresponds to common tip speed ratios of full scale turbines. The blades had a twist of 31° from root to tip and a pitch angle of 6° set at blade tip with respect to rotor plane. The hub height of the turbine is at 1.25D from the floor of the test section. Inside of the nacelle, a small direct current (DC) motor (*Faulhaber 1331T006SR*) is mounted, which is operated as a generator. The power output of the turbine was measured mechanically, therefore the torque T is measured by a sensor based on strain gauges with a design following the concept of the torque sensor of Kang *et al.* [32]. The rotational frequency ω is given by a magnetic encoder which is attached to the motor (*Faulhaber IE2-400*). The output of the encoder are 400 pulses per revolution. The measurements are performed using a National Instruments A/D converter of model *NI USB 6211* and in-house software written in *LabView*. The turbine was loaded by resistors to operate at its highest efficiency. The turbine is mounted in an gimbal support which allows oscillations in streamwise direction.

The turbine is stabilized by a cylindrical weight of 650 g below the gimbal. The pitch motion was characterized using video analysis. The analysis consisted of 300 s video recorded at 30 Hz of the floating turbine. The nacelle of the turbine was marked with a dot of high contrast and a video tracking tool was used to create a time series of the nacelle movement. The resulting mean pitch angle for the floating case is 17.6° with a standard deviation of 0.4°. The range of oscillations was between 16° and 19°. A Fourier transformation of the time series indicates that the dominant frequencies of the oscillation are in the range of 1.2–1.8 Hz. A range for the frequency has to be given since the oscillations of the turbine are not controlled and are induced by the inflow of the turbine. In part due to the lower rotor frontal area in the inclined turbine position, the power coefficient, c_p, in the floating case was 0.26 whereas a higher c_p of 0.29 was observed for the fixed case. Thrust coefficients, c_T, were estimated from the induction factor and found to be 0.89 and 0.85 for the fixed and floating cases, respectively. Induction factors were, in turn, calculated from the corresponding power coefficients [33]. Motion of the model was restricted to one-dimensional pitching oscillations, which is indeed the dominant motion in real conditions. Therefore, it is recognized that some aspects of the wake may differ from that of full-scale as sway and heave are not considered. Furthermore, in studies capturing the structural response of floating bodies to wave excitations as in offshore structures, the Froude number is taken into consideration [34]. In this study, given that the movement of the mast is restricted to pitch, Froude number is not considered. Scaled parameters for the model wind turbine that are taken into account are the tip speed ratio, thrust and power coefficient. The gimbal can be blocked at a straight position, so the turbine is operated as classic non-floating turbine and the performance of the turbine and surrounding flow conditions can be compared between fixed and floating with minor changes in the setup. Measurements were taken first for the fixed case, followed directly by measurements for the floating case.

3.3. SPIV

The SPIV setup consisted of a *LaVision* system with a Nd:YAG (532 nm, 1200 mJ, 4 ns duration) double pulsed laser and two 2 k × 2 k pixel CCD cameras. The time delay between exposures of the cameras was 150 μs. The seeding fluid was neutrally buoyant in air atomized diethylhexyl sebecate. To allow consistent resolution, the seeding density was kept constant during the measurements and seeding particles were well mixed within the wind tunnel. The thickness of the laser sheet was approximately 1 mm throughout the measurement plane. Before each experiment, the cameras were calibrated using a standard *LaVision* two-plane measurement plate with known geometries which are recognized by the *DaVis* measurement software. The resulting measurement plane was approximately 0.2 m × 0.2 m with a vector resolution of approximately 1.5 mm. To estimate the vector fields from the raw images, a multi-pass FFT based correlation algorithm was used. Interrogation windows of 64 × 64 pixels was used twice and a 32 × 32 window was used once, each with an overlap of 50%.

Figure 2. Wind turbine model with gimbal support. The gimbal is blocked for fixed case measurements. Scale 1:400, with $D = 0.2$ m. Power coefficient $c_p = 0.29$ for fixed case and 0.26 for floating case. Tip speed ratio $\lambda \approx 6$. Thrust coefficient $c_T \approx 0.89$ for the fixed and 0.85 for the floating case. Black coverage is to avoid laser beam reflections.

The freestream inflow conditions without the turbine present were measured using SPIV $1.5D$ to $0.5D$ upstream of the turbine position at a height of $-0.5D$ to $0.4D$ with $0D$ being hub height. For the wake measurements, the SPIV data was collected directly downstream of the centerline of the turbine. The height of the planes was positioned from $-0.25D$ to $0.75D$, so the averaged development of the blade tip vortices was captured. The planes were taken at distances of $0.7D$, $1.6D$, $2.5D$, $4.3D$ and $6D$. The first three planes were obtained with an overlap of approximately $0.1D$ to ensure a continuous plane throughout the near wake. SPIV allows for measurements of three velocity components in a two dimensional plane, where U is streamwise wind speed in the x direction, V the wall normal wind speed in the y direction and W is the spanwise wind speed in the z direction, as denoted in Figure 1.

For each measurement plane, 2500 samples were taken for the fixed and floating case. A convergence test was carried out by calculating ensemble averages of 500, 1500 and 2500 samples. The results between 1500 and 2500 samples match very well so statistical convergence of means and higher order moments is assured. Spurious vectors were excluded from statistical calculation using a normalized median test according to [35]. For all planes, the percentage of spurious vectors for U and V were below 1% and below 1.2% for W.

This work compares the wake development of the fixed case and the floating case. The fixed case represents the wake of a classical bottom-fixed turbine. The floating case represents the same turbine which has the freedom to incline in the streamwise direction (pitch motion). Due to the mean inflow velocity and its fluctuations, the turbine pitches downstream and oscillates, as described earlier in this work. Therefore, only the influence of dynamic pitch on the wake is discussed.

For the fixed turbine, the sampling frequency for the SPIV system of 1 Hz was used. In the floating case, the image acquisition was triggered to a fixed amplitude of oscillations in the downstream direction to ensure the same influence of the oscillation on wake structures for each image. Therefore, a reflective strip was placed on the tower of the oscillating turbine. A *Monarch* optical tracker was positioned in such way, that every time the turbine reached a fixed pitch angle the SPIV system was triggered for data collection. The angle was chosen to be close to maximum pitch angle ($\sim 18.5°$).

4. RESULTS

Measurements of the flow field obtained via SPIV for the fixed and floating turbine cases are compared. Contour plots are presented pairwise where the upper plot (Figure 3a) shows the result of the fixed turbine and the lower plot

(Figure 3b) shows the result of the floating turbine. The planes from 0.6D to 3.3D were taken with an geometrical overlap so the results of these planes are merged using a linear weight function. Thereafter, average profiles for the mean velocity, Reynolds stresses and terms in the mean kinetic energy equation are analyzed. Profile plots, which represent intersections of the wake at different downstream distances, are presented to emphasize the quantitative differences in the development of the wake.

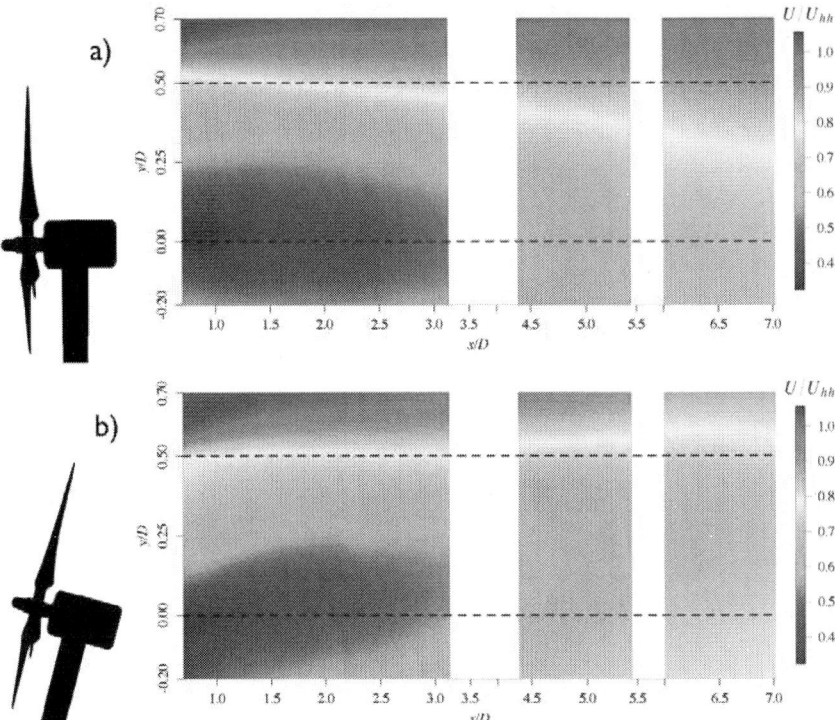

Figure 3. Normalized streamwise velocity component U/U_{hh} of the wake for the fixed and floating cases. U/U_{hh} in the fixed case has a typical symmetric shape around hub height. In the floating case it has a pronounced upwards trend with increasing x/D. Hub height ($y/D = 0$) and top blade tip ($y/D = 0.5$) are indicated by dashed horizontal lines.

4.1. Mean Flow

Figure 3 shows contour plots of the averaged streamwise velocity component U normalized by the inflow wind speed at hub height $U_{hh} = 6$ m/s. The velocity ranges are similar in both cases. In Figure 3a, U/U_{hh} behind the fixed turbine is plotted for a range of downstream distances encompassing both the near and far wake. Similarly, this is done in Figure 3b for the floating turbine, where the SPIV system is locked to a fixed pitch angle of the turbine. This angle represents a strong downstream pitching of the turbine due to its oscillations.

Figure 3a shows the wake of the fixed turbine which has the typical shape of a wake of a classical horizontal axis wind turbine [36,37]. The shape of the deficit is symmetric around hub height. From $-0.2D$ to $0.25D$ in the y/D dimension, a strong velocity deficit ($\sim$30% of U_{hh}) is observed due to the blockage of the nacelle. The deficit is pronounced mainly up to $3D$ and recovers up to 75% of U_{hh} at $7D$ downstream. These results correspond to the findings of Chamorro *et al.* [37] for a single turbine in an neutral boundary layer.

Figure 3b shows that, for the floating case, the deficit is no longer symmetric about the hub height in the wall normal direction, y/D. This is in stark contrast with the symmetry observed in the fixed case. The deficit is shifted down by $0.1D$, due to the pitch angle of the turbine. The area with the strongest deficit (below $0.5\ U_{hh}$) is by 25% smaller in the floating case compared to the fixed case. The wake pattern of the floating turbine is skewed upward by approximately 3° due to the pitch motion of the turbine. Similar wake deflections, due to non-axial momentum extraction of the turbine, were observed for turbines in yaw conditions, but in the spanwise direction [38,39]. The flow pattern in the shear layer above blade tip is more compact in the floating case than in the fixed case. At 6.5–$7D$ downstream, the low speed area (blue/green) is shifted up by approximately $0.25D$ towards the top tip. In addition, due to the overall upward shift of the wake, Figure 3b shows that the mean streamwise wind speed between the hub height and blade tip is 10% lower with $0.7\ U_{hh}$ in the floating case by 6.5–$7D$.

Figure 4 provides vertical profiles of U/U_{hh} at downstream distances x/D of $0.75D$, $1.5D$ and $3D$ for both the fixed and floating cases in order to allow a more detailed comparison. A logarithmic fit to the inflow profile is added to visualize the effect of the turbines on the flow. The fit extends the measured inflow profile to $y/D = 0.7$, since the measurement plane for the inflow was set to y/D of $-0.5D$ to $0.5D$. At x/D of $0.75D$, the shift of the deficit in the floating case is very pronounced. At x/D of $1.5D$, the profiles for both cases have similar shapes, but the wind speed is lower in the fixed case. At x/D of $3D$, a strong upwards shift by $0.1D$ in the y/D dimension is observed in the profile for the fixed case. In addition, the deficit is again weaker in the floating case than the fixed case. Comparing the fixed and floating cases, two main differences exist in the location of the inflection points and profile crossings near the top tip. First, the inflection point and crossings near the top tip are shifted vertically away from the tunnel floor for the fixed case with profile crossings located at $y/D =$

0.57D for the fixed case and at y/D = 0.55D for the floating case. Second, the magnitude of the deficit at which the profile crossing is located is smaller in the fixed case. The profiles for the fixed turbine tend to intersect at the lowest point whereas this does not occur for the floating turbine. This is attributed to the movement induced via the pitch motion of the turbine.

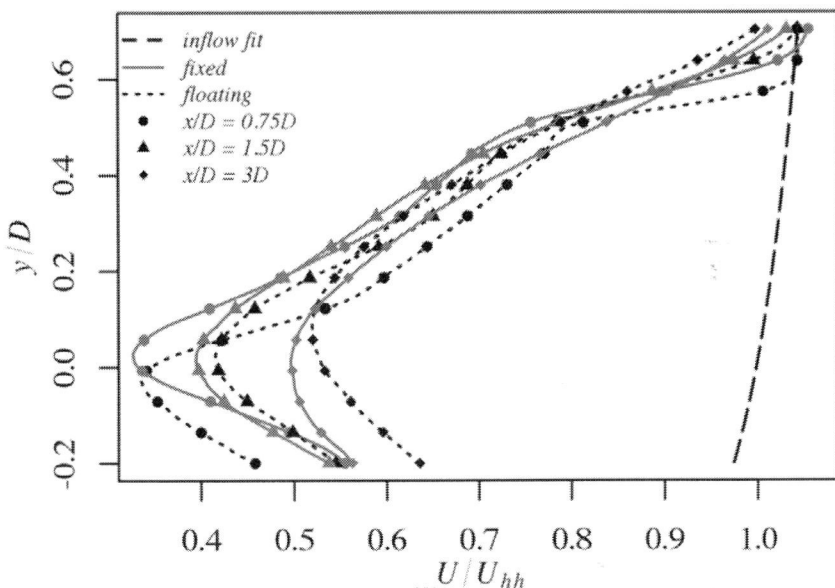

Figure 4. Profile of a logarithmic fit to inflow and near wake profiles of U/U_{hh} for the fixed and floating case at downstream distances 0.75D, 1.5D and 3D.

Figure 5 shows profiles at 4.5D, 6.1D and 7D downstream positions. The streamwise velocity component at blade tip height and above is smaller by 5%–15% for the fixed and by 10%–25% for the floating case as compared with U_{hh}, which is due the mast of the turbine being rigid and, consequently, further enhances mixing and recovery particularly in the shear layer of the wake. In the fixed case, the flow is recovering at hub height gradually, but has not fully recovered by 5–7D, which compares well to [37,40]. The profiles of the floating case are shifted up by 0.2–0.3D. In order to quantify the differences between the fixed and floating cases, the resultant thrust force (thrust force ~ U^2) from y/D = −0.2D to 0.5D was calculated. In the floating case, the thrust force on a downwind turbine is reduced 9%–10% compared to the fixed case. Also, the available power at downstream distances x/D = 4.5D, 6.1D and 7D is compared, by integrating the cubed velocity profiles from $y/D = -0.2D$ to $0.5D$. The available power to be extracted in the floating case is 14%–16% lower compared to the fixed case.

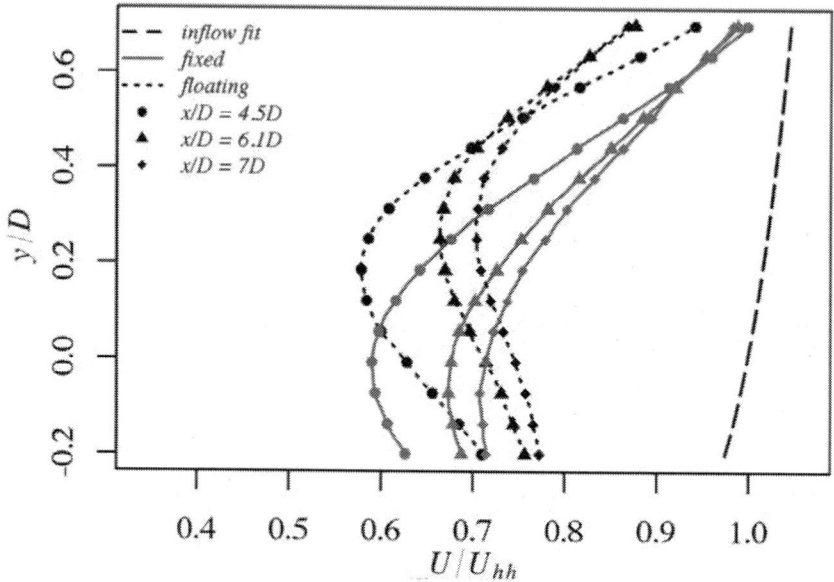

Figure 5. Profile of a logarithmic fit to inflow and far wake profiles of U/U_{hh} for the fixed and floating case at downstream distances $4.5D$, $6.1D$ and $7D$.

Figure 6 presents contours of mean wall normal velocity V/U_{hh} for both cases. In the fixed case (Figure 6a), the average wind speed hovers around 0 m/s, except directly behind the nacelle at hub height and below, where an area with $-0.04\ U_{hh}$ is present, thus entraining fluid downwards on the downstream range of 1–3D. As expected, the flow becomes increasingly homogeneous as x/D increases.

In the floating case (Figure 6b), the average wall normal velocity of the whole field is $0.035\ U_{hh}$. A small enclosed deficit area behind the nacelle with negative wind speed is present up to $1D$ downstream. A positive wind speed up to $0.08\ U_{hh}$ is shown from $y/D = 0.25$ and upwards toward the blade top tip. With increasing downstream distance, the wall normal velocity increases to an average of $0.04\ U_{hh}$ in the region from $6.1D$ to $7D$. Notably, a wall normal velocity of $0.04\ U_{hh}$ for the floating case represents 980% of V/U_{hh} for the same location in the fixed case.

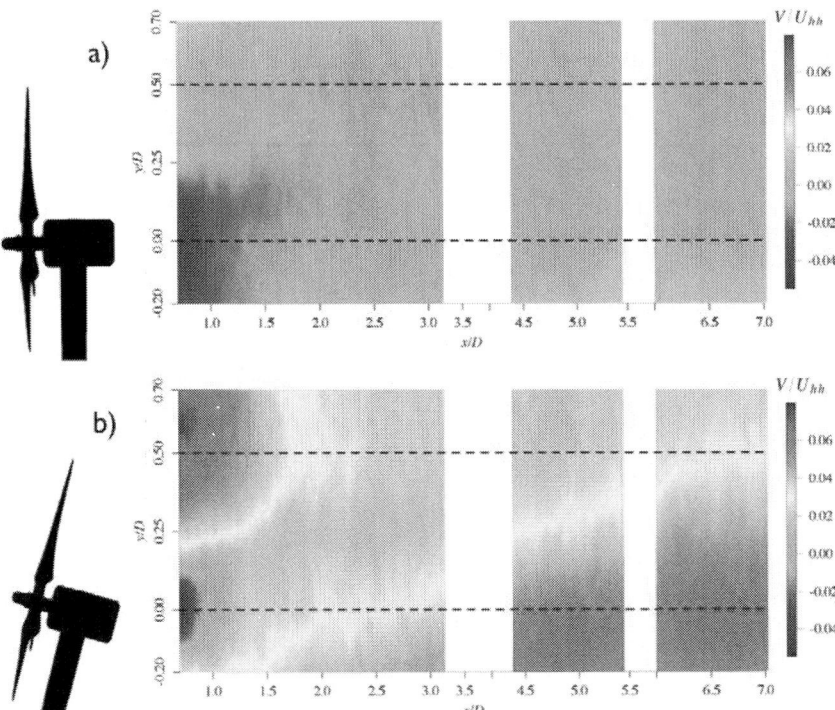

Figure 6. Normalized wall normal velocity component V/U_{hh} of the wake for the fixed and floating cases. In the fixed case, V/U_{hh} is close to zero. In the floating case, V/U_{hh} increases with increasing x/D. Hub height ($y/D = 0$) and top blade tip ($y/D = 0.5$) are indicated by dashed horizontal lines.

Profiles of V/U_{hh} at downstream distances x/D of $0.75D$, $1.5D$ and $3D$ are shown in Figure 7 for both cases. Above hub height, the profile shapes at x/D of $0.75D$ are similar for the two cases but the absolute velocities are much higher for the floating case. A well defined area of reversed flow behind the nacelle at hub height is evident in the wall normal velocity profile at $x/D = 0.75D$ for the floating case. In contrast, the area of reversed flow in the corresponding location is more diffused in the fixed case. A faster recovery is observed in the floating case than the fixed case.

Figure 8 presents far wake profiles for the wall normal velocity component at x/D of $4.5D$, $6.1D$ and $7D$ downstream. For the fixed case, not only are the trends in the velocity profiles at all three streamwise positions similar, but the absolute velocities observed are comparable. As observed in the wall normal velocity contour plots, the overall velocity is larger in the floating case.

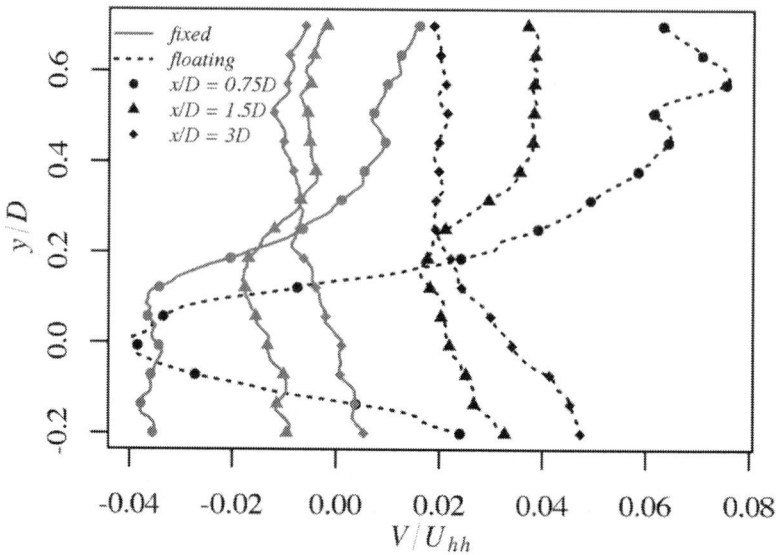

Figure 7. Near wake profiles of V/U_{hh} for the fixed and floating cases at downstream distances 0.75D, 1.5D and 3D.

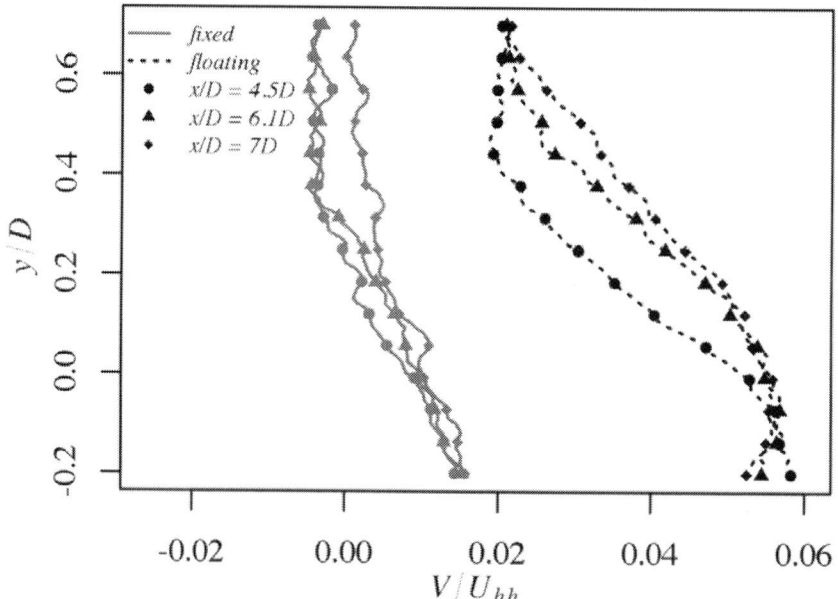

Figure 8. Far wake profiles of V/U_{hh} for the fixed and floating cases at downstream distances 4.5D, 6.1D and 7D.

Figure 9 presents contours of the mean spanwise velocity W/U_{hh}. The velocity ranges of both contours are comparable in magnitude although the shape of the contours differ from one another. In both cases, there is a clear visible divide between positive and negative spanwise velocity at hub height, which is due to the clockwise rotation of the blades. In the floating case, this point is shifted upwards by $0.03D$ due to the inclined turbine as a consequence of the oscillatory motion of the mast. In the fixed case, the area with negative velocity is cone-shaped and extends to $3D$ downstream of the rotor. In the fixed case, the line dividing areas of positive and negative velocities is roughly horizontal while the line dividing positive and negative velocities shows a positive slope with increasing downstream distance in the floating case.

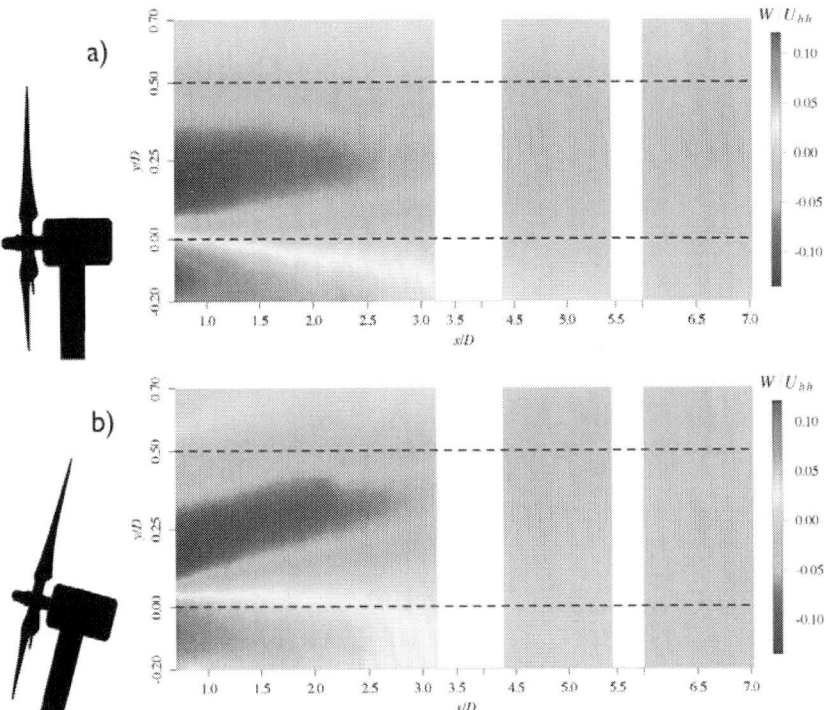

Figure 9. Normalized spanwise velocity component W/U_{hh} of the wake for the fixed and floating cases. In the fixed case, W/U_{hh} has a symmetric divide of positive and negative velocities. In the floating case, the negative shape moves upwards with increasing x/D, while the shape of contours of positive velocity stays at a constant height. Hub height ($y/D = 0$) and top blade tip ($y/D = 0.5$) are indicated by dashed horizontal lines.

Near wake profiles of spanwise velocity are presented in Figure 10. The profiles at $x/D = 0.75D$ have a similar shape for both cases, but the profiles develop differently with increasing distance. At $x/D = 1.5D$ and 3D, the shift of the peak negative speed in the floating case to higher wall normal position becomes evident. At 1.5D, the velocity magnitudes are shifted towards negative speeds in the floating case. The shape of the fixed profiles is symmetric, while the profiles in the floating case smear out with distance due to the oscillation of the turbine.

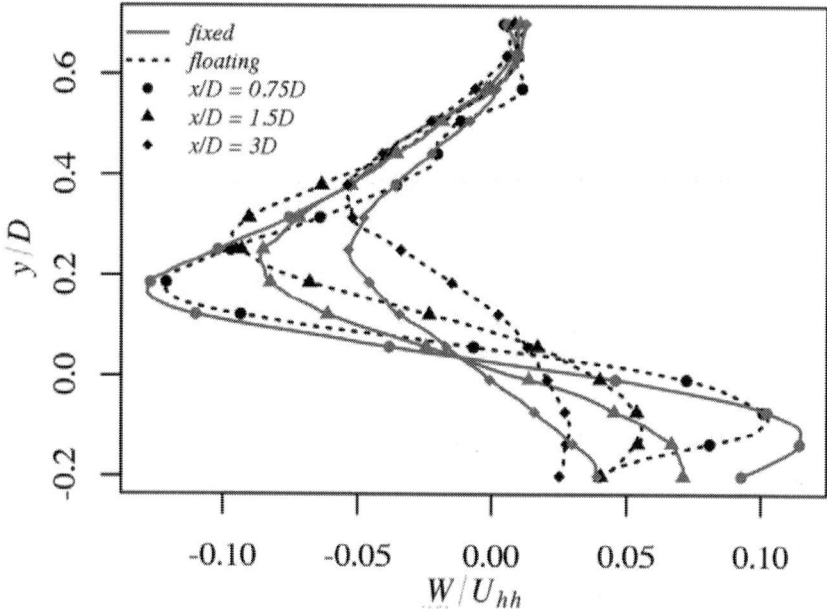

Figure 10. Near wake profiles of W/U_{hh} for the fixed and floating cases at downstream distances 0.75D, 1.5D and 3D.

Figure 11 shows the far wake profiles of W/U_{hh}. The minimum spanwise velocity occurs at 4.5D in the wall normal dimension for the fixed case whereas the minimum is shifted upward 0.2D in the wall normal dimension for the floating case. At $x/D = 6-7D$, the flow in the floating case is homogeneous and closer to zero, whereas in the fixed case the flow is negative from hub height to the tip of the blades and positive below hub height and above the tip. Such observations at $x/D = 6-7D$ suggest that the flow shedding due to the rotation of the blades is persisting further downstream in the fixed case even at 7D downstream.

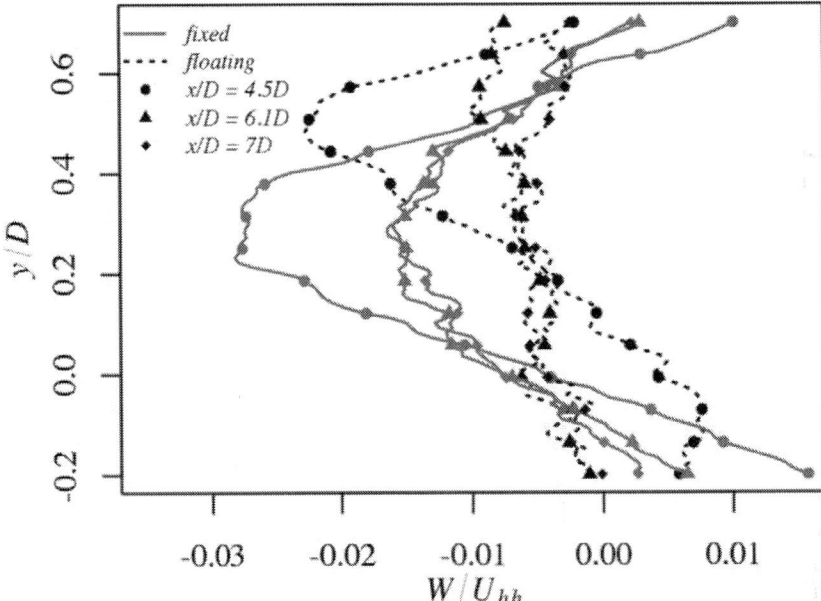

Figure 11. Far wake profiles of W/U_{hh} for the fixed and floating cases at downstream distances 4.5D, 6.1D and 7D.

4.2. Reynolds Stresses and Flux of Mean Kinetic Energy

Figure 12 shows the contours of the normalized normal component $\overline{uu}/U_{hh}^2$ of the Reynolds stress tensor. Overall, the average stress over the measurement area is 8% smaller in the floating case compared to the fixed case. In the fixed case, the largest magnitude of $\overline{uu}/U_{hh}^2$ is produced in the shear layer above the top tip. Directly behind the hub at hub height, roughly v-shaped regions with elevated magnitudes of $\overline{uu}/U_{hh}^2$ are evident for both cases. However, smaller magnitudes of the Reynolds stress are found in this v-shaped region in the fixed case. Furthermore, in the non-moving turbine, these features diminish after $x/D =$ 1.5–2D. In the case of the oscillating turbine, the presence of a higher stress behind and above hub height is advected towards the shear layer and at this wall-normal location with a downstream distance of $x/D \approx 3$, it merges. In the near wake of the floating case, the magnitudes of $\overline{uu}/U_{hh}^2$ in the shear layer as well as behind the hub are comparable to the magnitudes of the fixed case.

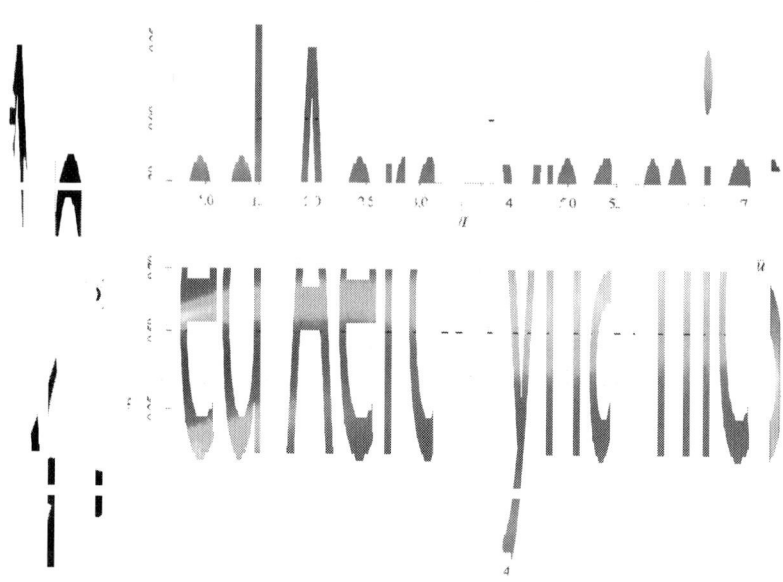

Figure 12. Contour of the normalized $\overline{uu}/U_{hh}^2$ Reynolds stress term for the fixed and floating cases. In the fixed case, most of the stress is created above blade tip and extends downward with increasing x/D. In the floating case, high $\overline{uu}/U_{hh}^2$ is created behind the hub and above tip top. Hub height ($y/D = 0$) and top blade tip ($y/D = 0.5$) are indicated by dashed horizontal lines.

Above the top tip, $\overline{uu}/U_{hh}^2$ in the shear layer is 20% higher in the fixed case for downstream distances up to $x/D \approx 3.1D$ and, in the far wake, the stress in the upper segment ($y/D > 0.5$) is 26% higher in the floating case. The maximum amplitude of $\overline{uu}/U_{hh}^2 = 0.023$ is at $y/D \approx 0.57D$ in the fixed case.

In the floating case, the peak above top tip is lower with $\overline{uu}/U_{hh}^2 = 0.019$ at $y/D \approx 0.54D$ whereas the peak below hub height is the highest streamwise Reynolds normal stress with $\overline{uu}/U_{hh}^2 = 0.025$ at $y/D \approx -0.18D$.

In Figure 13, the Reynolds shear stress, $\overline{uv}/U_{hh}^2$, is presented. This quantity is an indicator for the transport of momentum. The development of $\overline{uv}/U_{hh}^2$ is of high interest as it is responsible for the energy being extracted from the turbine [16]. In both cases, the ranges of turbulent shear stress are close in magnitude.

The range and shape of the contours in the fixed case is comparable to the results found by Chamorro *et al.* [37]. A large negative turbulent stress above hub height and a positive turbulent stress below hub height is present, which is attributed to mixing effects of the wake.

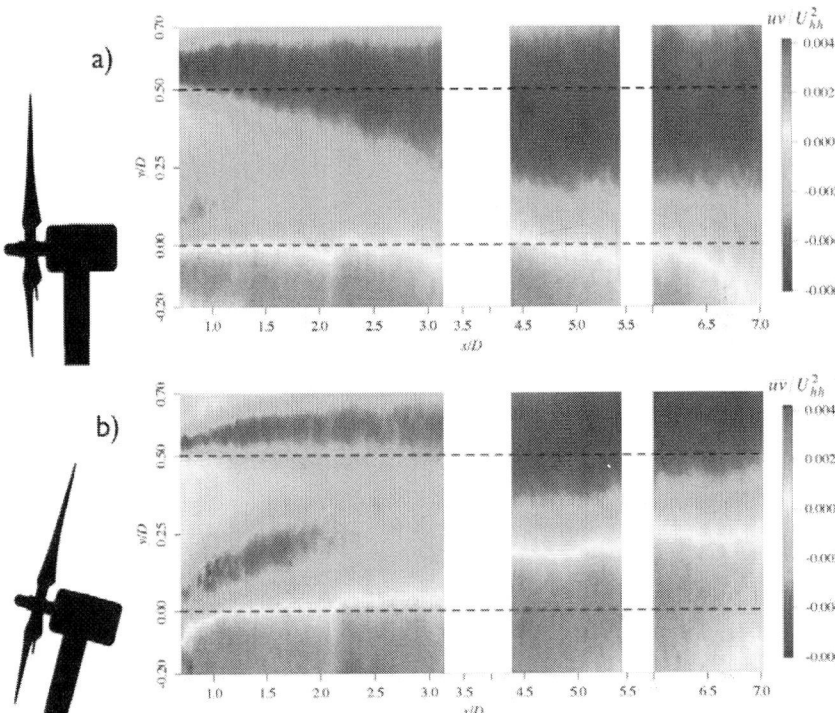

Figure 13. Contours of Reynolds shear stress $\overline{uv}/U_{hh}^2$ for the fixed and floating cases. Hub height ($y/D = 0$) and top blade tip ($y/D = 0.5$) are indicated by dashed horizontal lines.

In the fixed case, positive $\overline{uv}/U_{hh}^2$ is mostly present below hub height. In the floating case, the oscillations of the turbine cause a stronger positive shear stress in the far wake. Positive shear stress becomes less significant in the far wake region, $x/D \approx 6.1$–$7D$, in the fixed case. In the floating case, positive shear stress dominates in this same far wake region. Furthermore, the positive Reynolds shear stress region below hub height remains constant in magnitude with x/D for the fixed case whereas this region is shifted upwards in the floating case. The pitch motion of the turbine causes a larger variation of the shear stress in the floating case and therefore stronger changes in the momentum flux are observed.

Figure 14 shows the near wake profiles of the Reynolds shear stress $\overline{uv}/U_{hh}^2$. For the fixed case, the profiles cross just above the hub height whereas in the floating case they tend to monotonically increase with downstream position. Right behind the rotor ($x/D = 0.75$), the Reynolds shear stress magnitudes are also greater for the fixed turbine.

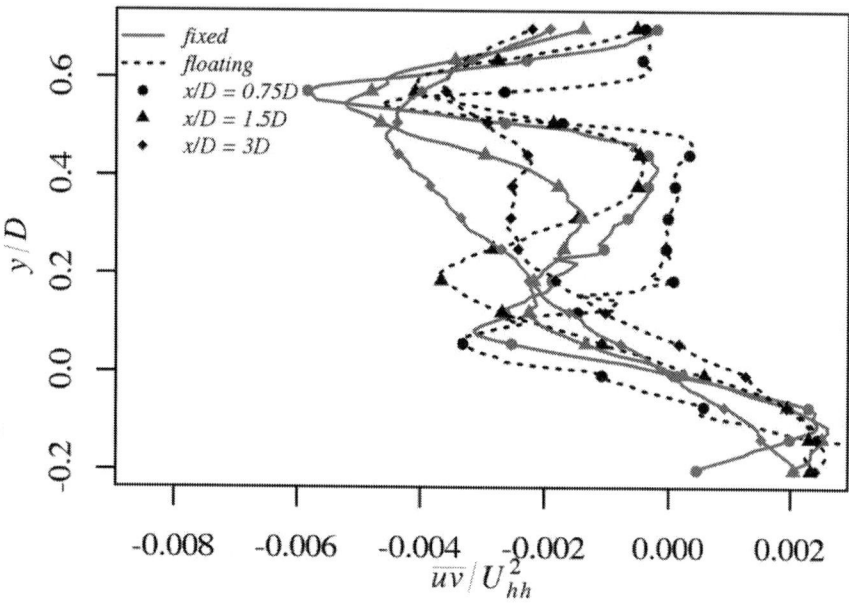

Figure 14. Near wake profiles of $\overline{uv}/U_{hh}^2$ for the fixed and floating cases at downstream distances 0.75D, 1.5D and 3D.

Figure 15 shows far wake profiles of $\overline{uv}/U_{hh}^2$ for fixed and floating cases. Although the shapes of the profiles for both case evolve similarly with increasing distance downstream, there is a systematic upward shift in the profiles for the floating case. The negative peaks of the stress at 6.1D and 7D are similar in magnitude in both cases.

Figure 16 shows the contours of *TKE*, which has shapes that develop similar to the normal stress component $\overline{uu}/U_{hh}^2$ in Figure 12. This quantity is important, since for the first turbine, the convective terms in the turbulent kinetic energy equation are significant. For both cases, turbulent kinetic energy has high values above top tip and and behind the hub. The maximum magnitudes in the floating case are higher, especially behind the hub, where *TKE* has its maximum with 1.4 m²/s², while for the fixed case, the highest *TKE* is in the shear layer with 0.7

m²/s². For the fixed case, the turbulent kinetic energy from above tip top spreads and moves downward with increasing x/D with decaying magnitude in the near wake and increases from $4.5 < x/D < 5.5$. In the floating case, the magnitude of *TKE* behind the hub decreases quickly with increasing downstream distance $x/D = 0.7$–2.2. The *TKE* above hub height spreads slightly with increasing x/D, but the structure remains close to the top tip height and moves upwards in the far wake.

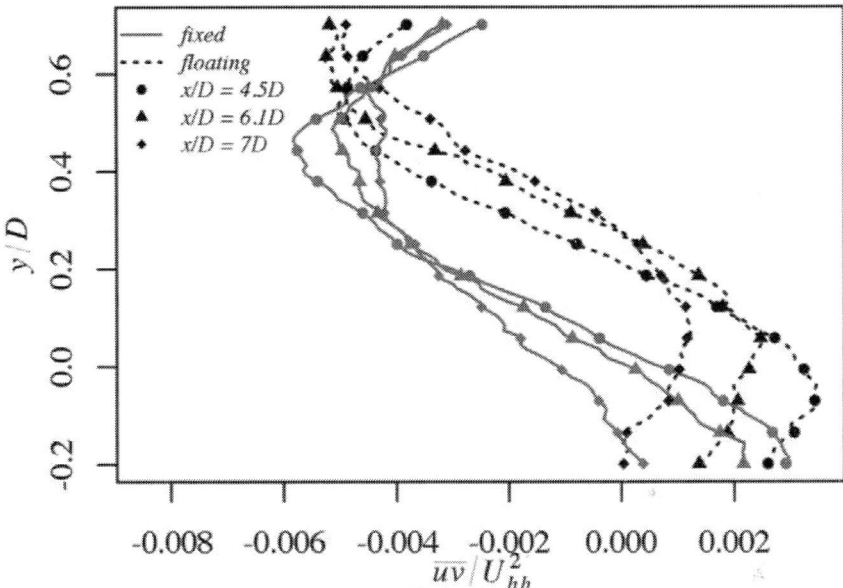

Figure 15. Far wake profiles of $\overline{uv}/U_{hh}^2$ for the fixed and floating cases at downstream distances $4.5D$, $6.1D$ and $7D$.

Figure 17 shows the development of the flux of the Reynolds shear stress $-\overline{uv}U$, which has been found to be important for the energy being extracted by the turbine [16]. For $y/D < 0$, $-\overline{uv}U$ is negative for both cases. In the floating case, this region shifts away from the wall with increasing downstream distance. In the fixed case, a positive flux is observed in the shear layer and the quantity grows over a larger area with increasing downstream distance. At the top tip of the rotor, the magnitude remains relatively constant at about 0.9 m³/s³. In contrast, in the floating case, the area of positive flux increases more slowly with increasing distance. Also at the top tip of the rotor, the flux of mean kinetic energy increases in magnitude as the flow advects downstream for the floating case as well as moving away from the top tip rotor location. The large negative blue area is restricted to $y/D < 0$ on the fixed case. This is contrary to the floating case where the flux crosses the $y/D < 0$ into the top half of the rotor and it is permanent even at $x/D = 7D$.

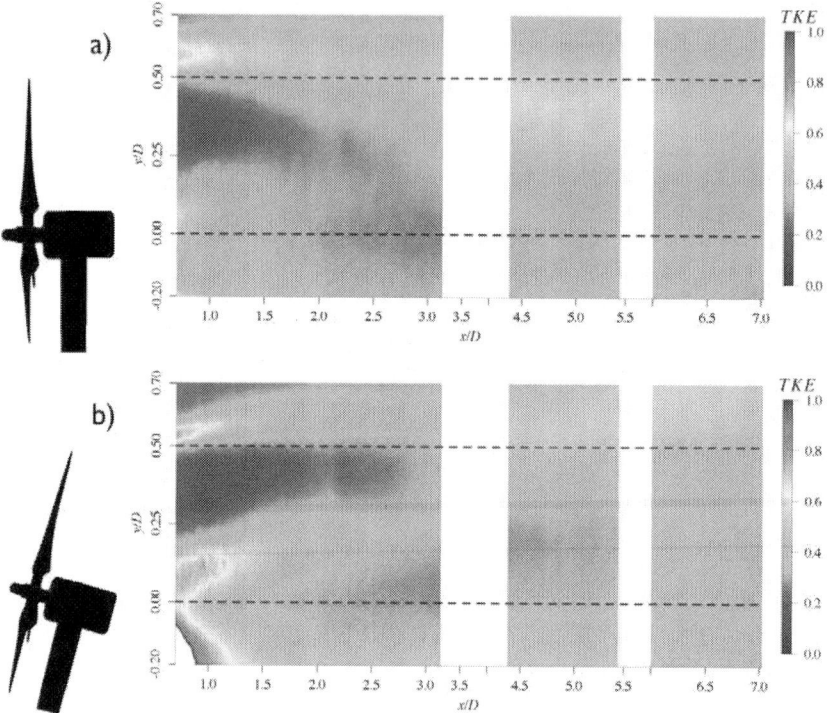

Figure 16. Contours of turbulent kinetic energy $TKE = \frac{1}{2}(\overline{uu} + \overline{vv} + \overline{ww})$ for (a) fixed and (b) floating case. The shapes develop analog to $\overline{uu}/U_{hh}^2$. Hub height ($y/D = 0$) and top blade tip ($y/D = 0.5$) are indicated by dashed horizontal lines.

The contours of vertical flux of this normal Reynolds stress component $-\overline{vv}V$ for the fixed and floating cases in Figure 18a,b, respectively. For the fixed case, the largest magnitudes of the mean vertical flux exist behind the hub, but the overall mean vertical flux of $-\overline{vv}V$ is close to zero with -0.004 m³/s³. For the floating case, $-\overline{vv}V$ is mostly positive with an average value of 0.04 m³/s³. Although these components are small compared to $-\overline{uv}U$, a remarkable difference exists between the two cases, which is consequently attributed to the oscillations of the turbine, thus contributing to enhanced transport due to the vertical fluctuations.

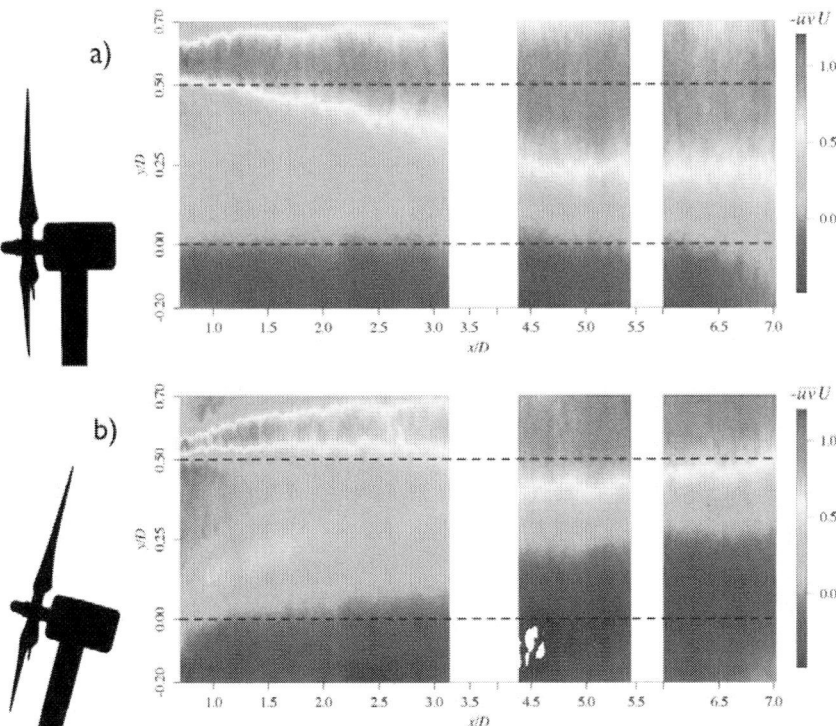

Figure 17. Contours for the flux of Reynolds shear stress $-\overline{uv}U$ for fixed (**a**) and floating (**b**) case. The flux of the shear stress represents the power that can be extracted. Hub height ($y/D = 0$) and top blade tip ($y/D = 0.5$) are indicated by dashed horizontal lines.

4.3. Comparison with Models

Measured profiles of U/U_{hh} are compared to wake models as proposed by [19,20,24,27–29,41] at positions of $x/D = 1.5D$, $3D$, $4.5D$ and $7D$ and the models are described in Section 2.2. These are shown in Figures 19 and 20 for the fixed and floating turbine cases, respectively.

For the fixed case, all wake models with the exception of the Larsen model overestimate the wake expansion at $1.5D$. For the Jensen model, which has a hat-shaped profile, the output of the model does not match the experimental data. Since the prescription of the model is solely determined by k in the far wake behavior, this difference is expected. The Ainslie model is set to a constant Gaussian profile for distances less than $2D$ and hence at $x/D = 1.5D$, the deficit is greater than the experimental data. The uniform actuator disk RANS model results in a wake that is comparable to the diameter of the disk. This disk acts like a porous bluff body, and ignores aerodynamic details like tip vortices, nacelle and tower effects, that causes, once again, the profile of the wake to be more affected by the rotor than the experimental data.

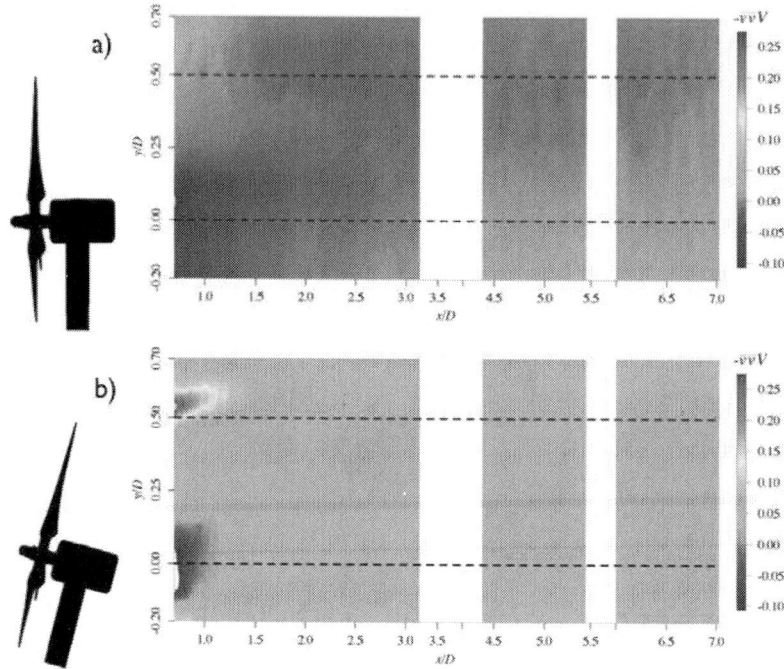

Figure 18. Turbulent kinetic flux in the wall normal direction component $-\overline{vv}V$ of the wake for fixed and floating cases.

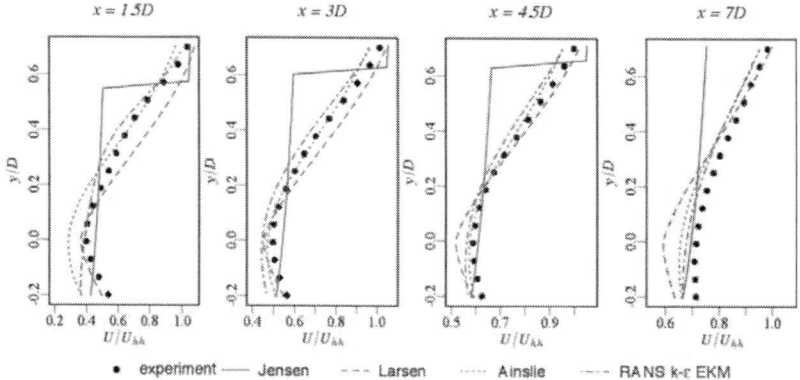

Figure 19. Comparison of measured mean profiles of U/U_{hh} with various wake models for the fixed case.

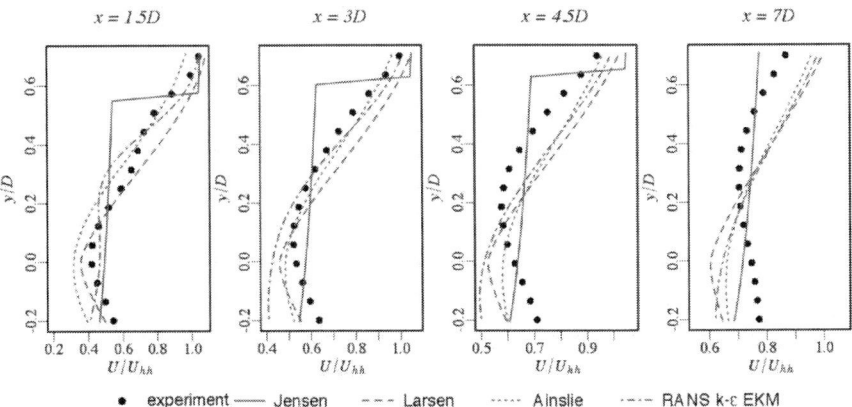

Figure 20. Comparison of measured mean profiles of U/U_{hh} with various wake models for the floating case.

At larger distances x/D of $4.5D$ and $7D$ location of the maximum of the predicted wake deficit coincides with the experimental data for the fixed case. Nevertheless, the rest of the profile does not collapse with the experimental data due to inherent shape of the profile (*i.e.*, top-hat shape). Furthermore, the Ainslie and RANS models improve considerably in comparison at these downstream distances although the maximum is slightly over-predicted. Taking into account all of the employed models, the RANS model best matches the experimental data as it takes into account additional terms compared with the other models. Of particular interest is the influence due to the wall, which is related to the influence of the developing boundary layer. Notably, the models tend to over-predict the velocity deficit for the fixed case when compared to the acquired data. This is due to the assumption built into the Jensen, Larsen and Ainslie models that the total wake deficit is calculated based on the concept of superposition of the inflow profile in addition to the modeled deficit. It is well understood that the inflow profile is not constant as it advects downstream. On the other hand, the RANS simulation captures the changes in the development of the wake and more realistically predicts the shape of the velocity deficit. However, an offset at all measured distances is clearly present, which can be attributed to either the parameterization of the wall and/or the turbulence closure. The latter inherently follows from the use of an eddy viscosity in the wake region that tends to be a larger sink that otherwise prescribed.

In stark contrast to what was observed for the fixed case, the models do not manage to capture the vertical shift to higher y/D of the measured wake of the floating case. Figure 20 shows that these discrepancies become more apparent as downstream distance increases. It is of no surprise that the velocity profiles generated via the Jensen, Larsen and Ainslie models do not collapse with the experimental data since these models have not been constructed to represent a wind turbine with induced pitch due to the incoming flow.

In fact, both the Jensen and Ainslie models assume an axial symmetry in the wake. As a result of the pitch motion not being well represented in the models, the wake is then directed towards the wall. Therefore, the overall effect is reduced to a change in both thrust and power coefficients. The RANS model was implemented differently than the other models in that the RANS model was tilted by 15° and the rotor position was shifted accordingly. Even though these changes are implemented, the RANS model still is not able to capture the behavior as that observed in the experimental data. This disparity is then attributed to the fact that rotation was not represented in the applied actuator disk model. Consequently, it cannot capture the anisotropy as observed by the wake.

5. DISCUSSION

Platform pitch and streamwise oscillations have a strong impact on the mean shape of the wake as well as the magnitudes of all velocity components. Due to the oscillations in the floating case, the turbine experiences a variable shear flow. The pitch motion of the platform with an average inclination angle of 17.6° skews the mean streamwise velocity component in the wake with a positive slope, with an angle of approximately 3°. This observation is comparable to results observed in yawed turbines with a yaw misalignment of 20° [38,39]. Extending the analogy of wake deflection from turbines under a yaw condition, the average skew angle of the wake may be partially explained by momentum theory since the non-axial momentum extraction is due to the pitching effect of the turbine [33]. In order to capture other effects generated by the oscillation of the platform, more complex descriptions are necessary as highlighted by Sebastian et al. [42]. The shift due to the pitch motion/oscillations results in a far wake, which has not fully recovered and thus 14%–16% less power is available for the downwind turbine. Furthermore, the 10% decrease in thrust force observed in the floating turbine is due to the decrease in mean velocity and upward shift of the wake. It is also expected that a completely different load distribution on the rotor of a downwind turbine is observed compared to that of non-floating turbine.

The increase of the wall-normal mean velocity in the floating case is of importance, since vertical flow is often not considered in wake models. This is certainly true considering the development of V/U_{hh} for the fixed case (Figure 6b), which subsequently, leads to potential, unpredicted fatigue loads on downstream turbines in the floating case. The shape of the V/U_{hh} wake component is not changed by the oscillations (compare contours in Figure 6a,b), but the magnitudes are increased, resulting in high vertical wind speeds in the floating case. It can then be concluded that the vertical component of the flow field in the wake of an floating turbine cannot be neglected and must be considered for operating conditions of downstream turbines.

When evaluating the out of plane component, that is the spanwise component of the mean velocity, W/U_{hh}, symmetry in this component between the top and bottom of the rotor (i.e., $y/D > 0$ and $y/D < 0$, respectively) for the fixed turbine is entirely disrupted when oscillations are taken into account as seen in the floating case. Although as the flow develops downstream, the signatures due to the out of plane velocity are less pronounced in comparison to the fixed case.

Furthermore, the oscillations and inclination experienced by the turbine have a strong impact on the fluctuations of the flow and on the features of turbulent kinetic energy in the wake, where in the top tip of the fixed case, the features of $\overline{uu}$, $\overline{uv}$ and TKE seem to diffuse with increasing distance. These features become more pronounced with increasing downstream distance for the oscillating turbine. The evaluation of the turbulent kinetic energy for the turbine without induced motion is maximum at the top tip of the blade and is shifted upwards in the oscillating case. At hub height, the value of the TKE nearly doubles again pointing towards the influence of the pitch dynamics. This increase can be linked with the increase in the mean vertical velocity component.

From the point of view of a downwind turbine, the shear stress distribution in the fixed case leads to negative fluctuating shear forces above hub height and positive fluctuating shear forces below the hub. Similarly, strong fluctuations are observed at the nacelle. In the floating case, most of a downwind rotor potentially experiences positive fluctuating shear forces, but a strong transition from positive to negative shear force occurs at the blade tips.

Turbulent kinetic energy is a measure for the amount of kinetic energy that is contained in the fluctuations. For the fixed case, the rotor of a downwind turbine would be exposed to strong fluctuations above hub height. In the floating case, the smaller TKE in the far wake results in a steadier inflow situation for a downstream turbine, which is positive from the point of view of loads on a turbine. The vertical shift of the turbulence quantities, on the other hand, affects the power production adversely. When considering the mean kinetic energy flux, $-\overline{uv}U$, the upwards shift as seen in the floating case results in a lower momentum entrainment from overhead flow, thus resulting in less energy available in the downstream wind and therefore in less power available for a downstream turbine.

Almost no vertical transport of vertical fluctuations can be observed in the fixed case, which compares well to findings of [30], where $-\overline{vv}V$ is several orders of magnitudes smaller than the main contributor to the flux $-\overline{uv}U$.. In the floating case, $-\overline{uv}U$ remains much bigger than $-\overline{vv}V$, but the pitch motion of the turbine creates high positive transport at the blade tips and a negative transport behind the hub. Due to the overall positive vertical flux, entrainment of kinetic energy from the free stream above the canopy is reduced in the floating case. This corresponds well to the flow in the mean streamwise direction, where the wake recovery is diminished as observed by the far wake, thus contains less kinetic energy for a downstream turbine.

The comparison with wake models shows that for the fixed case, the shape and magnitudes of the streamwise component of the wake in midrange distance ($x = 3$–$7D$) can be approximated by the Ainslie, the Larsen and the actuator disk models, even though in the lower half of the wake the models over-predict the deficit. In the floating case, all models fail to capture the vertical displacement of the streamwise component, resulting in an inaccurate wake description. The inflow of a downwind turbine would then be inaccurate leading to erroneous load and power predictions for a downwind turbine.

6. CONCLUSIONS

Wind tunnel experiments were performed to compare the wake development of a fixed and a streamwise oscillating wind turbine model using stereo PIV. Statistical analysis of wake development from $0.7D$ to $7D$ was performed and differences in their quantities were elaborated for both cases. The pitch motion of the turbine has a strong impact on the development of the mean components as well as on the turbulent quantities. The vertical shift of U/U_{hh} results in less available kinetic energy for a downwind turbine in offshore conditions, which could be due to the shifted turbulent kinetic flux $-\overline{uv}U$. The upwards shift of the turbulent kinetic energy reduces potential fatigue loads on a downwind turbine, but the increase in V/U_{hh} adds to such loads. Comparison of the measurements with wake models reveals minor discrepancies of the model predictions in the fixed case, but these models fail to describe the wake behind a wind turbine with induced dynamic pitch motion. Floating platforms for wind turbines provide the capability for allowing the production of energy offshore. Nevertheless, the pitch motion of the turbine results in new challenges for wake modeling, thus providing the opportunity for furthering the modeling capabilities to a new scenario (*i.e.*, offshore *versus* onshore). This certainly has great implications in the experienced loads by the turbine as well as the power production capabilities. The article examines the dynamic pitch motion of a wind turbine; it would also be relevant to discern the differences between the dynamic pitching with static pitch in the future.

ACKNOWLEDGMENTS

The authors would like to thank Stefan Ivanell for providing the blade design, Carlos Peralta for providing the turbulence model implementation for the actuator disk model simulation. This work was supported in part by grants from the Federal Environmental Foundation (DBU), Germany.

CONFLICTS OF INTEREST

The authors declare no conflicts of interest.

REFERENCES

1. Zervos, A.; Kjaer, C. *Pure Power: Wind Energy Scenarios up to 2030*; European Wind Energy Association: Brussels, Belgium, 2008.

2. Henderson, A.; Witcher, D.; Morgan, C. Floating Support Structures Enabling New Markets for Offshore Wind Energy. Proceedings of the European Wind Energy Conference (EWEC), Marseille, France, 16–19 March 2009.

3. Butterfield, C.; Musial, W.; Jonkman, J. *Engineering Challenges for Floating Offshore Wind Turbines*; National Renewable Energy Laboratory: Golden, CO, USA, 2007.

4. Sebastian, T.; Lackner, M. Analysis of the induction and wake evolution of an offshore floating wind turbine. *Energies***2012**, *5*, 968–1000.

5. Barthelmie, R.J.; Pryor, S.C.; Frandsen, S.T.; Hansen, K.S.; Schepers, J.G.; Rados, K.; Schlez, W.; Neubert, A.; Jensen, L.E.; Neckelmann, S. Quantifying the impact of wind turbine wakes on power output at offshore wind farms. *J. Atmos. Ocean. Technol.* **2010**, *27*, 1302–1317.

6. Musial, W.; Butterfield, S.; Renewable, N. Energy from Offshore Wind. Proceedings of the Offshore Technology Conference, Houston, TX, USA, 1–4 May 2006; pp. 18355:1–18355:11.

7. Jonkman, J. Dynamics of offshore floating wind turbines—Model development and verification. *Wind Energy* **2009**, *12*, 459–492.

8. Jonkman, J.; Matha, D. *A Quantitative Comparison of the Responses of Three Floating Platforms*; National Renewable Energy Laboratory: Golden, CO, USA, 2010.

9. Jonkman, J.; Matha, D. Dynamics of offshore floating wind turbines— Analysis of three concepts. *Wind Energy* **2011**,*14*, 557–569.

10. Matsukuma, H.; Utsunomiya, T. Motion analysis of a floating offshore wind turbine considering rotor-rotation. *IES J. Part A Civ. Struct. Eng.* **2008**, *1*, 268–279.

11. Utsunomiya, T.; Matsukuma, H.; Minoura, S. On Sea Experiment of a Hybrid SPAR for Floating Offshore Wind Turbine Using 1/10 Scale Model. Proceedings of the ASME 29th International Conference on Ocean, Offshore and Arctic Engineering, Shanghai, China, 6–11 June 2010; Volume 3, pp. 529–536.

12. Wang, C.M.; Utsunomiya, T.; Wee, S.C.; Choo, Y.S. Research on floating wind turbines: A literature survey. *IES J. Part A Civ. Struct. Eng.* **2010**, *3*, 267–277.

13. Matha, D.; Schlipf, M.; Cordle, A.; Pereira, R.; Jonkman, J. Challenges in Simulation of Aerodynamics, Hydrodynamics, and Mooring-Line Dynamics of Floating Offshore Wind Turbines. Proceedings of the 21st Offshore and Polar Engineering Conference, Maui, HI, USA, 19–24 June 2011.

14. Sebastian, T.; Lackner, M. Characterization of the unsteady aerodynamics of offshore floating wind turbines. *Wind Energy* **2013**, *16*, 339–352.

15. Barthelmie, R.J.; Larsen, G.C.; Frandsen, S.T.; Folkerts, L.; Rados, K.; Pryor, S.C.; Lange, B.; Schepers, G. Comparison of wake model simulations with offshore wind turbine wake profiles measured by sodar. *J. Atmos. Ocean. Technol.***2006**, *23*, 888–901.

16. Cal, R.B.; Lebrón, J.; Castillo, L.; Kang, H.S.; Meneveau, C. Experimental study of the horizontally averaged flow structure in a model wind-turbine array boundary layer. *J. Renew. Sustain. Energy* **2010**, *2*.

17. Calaf, M.; Meneveau, C.; Meyers, J. Large eddy simulation study of fully developed wind-turbine array boundary layers. *Phys. Fluids* **2010**, *22*.

18. Davidson, P.A. *Turbulence: An Introduction for Scientists and Engineers*; Oxford University Press: Oxford, UK; New York, NY, USA, 2004.

19. Ainslie, J.F. Development of an Eddy Viscosity Model for Wind Turbine Wakes. Proceedings of the 7th BWEA Wind Energy Conference, Oxford, UK, 23–25 March 1983; pp. 61–66.

20. Jensen, J.O. *A Note on Wind Generator Interaction*; Technical Report Risø-M-2411; Risø National Laboratory: Roskilde, Denmark, 1983.

21. Nygaard, N.G. Construction and Validation of a New Offshore Wake Model. Proceedings of the International Conference on Aerodynamics of Offshore Wind Energy Systems and Wakes (ICOWES) Conference, Lyngby, Danmark, 17–19 June 2013.

22. Schmidt, J.; Stoevesandt, B. Wind Farm Layout Optimisation Using Wakes from Computational Fluid Dynamics Simulations. Proceedings of the EWEA Conference, Barcelona, Spain, 10–13 March 2014.

23. Lange, B.; Waldl, H.P.; Guerrero, A.G.; Heinemann, D.; Barthelmie, R.J. Modelling of offshore wind turbine wakes with the wind farm program FLaP. *Wind Energy* **2003**, 6, 87–104.

24. OpenFOAM, 2013. Available online: http://www.openfoam.org (accessed on 5 November 2013).

25. Larsen, G.C. *A Simple Wake Calculation Procedure*; Technical Report Risø-M-2760; Risø National Laboratory: Roskilde, Denmark, 1988.

26. Swain, L.M. On the turbulent wake behind a body of revolution. *Proc. R. Soc. Lond. A* **1929**, 125, 647–659.

27. Manwell, J.F.; McGowan, J.G.; Rogers, A.L. *Wind Energy Explained*; Wiley and Sons: Chichester, UK, 2009.

28. El Kasmi, A.; Masson, C. An extended k-ϵ model for turbulent flow through horizontal-axis wind turbines. *J. Wind Eng. Ind. Aerodyn.* **2008**, 96, 103–122.

29. Richards, P.J.; Hoxey, R.P. Appropriate boundary conditions for computational wind engineering models using the k-ϵ turbulence model. *J. Wind Eng. Ind. Aerodyn.* **1993**, 46–47, 145–153.

30. Hamilton, N.; Melius, M.; Cal, R.B. Wind turbine boundary layer arrays for Cartesian and staggered configurations—Part I, flow field and power measurements. *Wind Energy 2014*.

31. Simms, D.; Schreck, S.; Hand, M.; Fingersh, L. *NREL Unsteady Aerodynamics Experiment in the NASA-Ames Wind Tunnel: A Comparison of Predictions to Measurements*; National Renewable Energy Laboratory: Golden, CO, USA, 2001.

32. Kang, H.S.; Meneveau, C. Direct mechanical torque sensor for model wind turbines. *Meas. Sci. Technol.* **2010**, *21.*

33. Burton, T.; Jenkins, N.; Sharpe, D.; Bossanyi, E. *Wind Energy Handbook*; John Wiley & Sons: Chichester, UK, 2011.

34. Robertson, A.; Jonkman, J.; Masciola, M. Summary of Conclusions and Recommendations Drawn from the DeepCWind Scaled Floating Offshore Wind System Test Campaign. Proceedings of the ASME 32nd International Conference on Ocean, Offshore and Arctic Engineering, Nantes, France, 9–14 June 2013; Volume 8.

35. Westerweel, J.; Scarano, F. Universal outlier detection for PIV data. *Exp. Fluids* **2005**, *39*, 1096–1100.

36. Vermeer, L.; Sø rensen, J.; Crespo, A. Wind turbine wake aerodynamics. *Prog. Aerosp. Sci.* **2003**, *39*, 467–510.

37. Chamorro, L.P.; Porté-Agel, F. Effects of thermal stability and incoming boundary-layer flow characteristics on wind-turbine wakes: A wind-tunnel study. *Bound.-Layer Meteorol.* **2010**, *136*, 515–533.

38. Medici, D. Experimental Studies of Wind Turbine Wakes Power Optimisation and Meandering. Ph.D. Thesis, Royal Institute of Technology, Stockholm, Sweden, December 2005.

39. Parkin, P.; Holm, R.; Medici, D. The Application of PIV to the Wake of a Wind Turbine in Yaw. Proceedings of the Particle Image Velocimetry, Gottingen, Germany, 17–19 September 2001; pp. 155–162.

40. Wu, Y.T.; Porté-Agel, F. Large-eddy simulation of wind-turbine wakes: Evaluation of turbine parametrisations.*Bound.-Layer Meteorol.* **2010**, *138*, 345–366.

41. Larsen, G.C. *A Simple Stationary Semi-Analytical Wake Model*; Technical Report Risø-R-1713; Risø National Laboratory: Roskilde, Denmark, 2009.

42. Sebastian, T.; Lackner, M.A. Development of a free vortex wake method code for offshore floating wind turbines.*Renew. Energy* **2012**, *46*, 269–275.

43. Renkema, D.J. Validation of Wind Turbine Wake Models. Master's Thesis, Delft University of Technology, Delft, The Netherlands, June 2007.

CHAPTER 3

Sport Aerodynamics: on the Relevance of Aerodynamic Force Modelling Versus Wind Tunnel Testing

Caroline Barelle[1]

[1] *National Technical University of Athens, Greece*

1. INTRODUCTION

In sports events, performance analysis is not an easy task since multiple factors, such as physiology, psychology, biomechanics, and technical progress in equipment are simultaneously involved and determine the final and ultimate outcome. Identification of individual effects are thus complicated, however from a general point of view, aerodynamics properties are recognized to play a determinant role in almost every sports in which the performance is the result of the optimal motion of the athlete (multi-jointed mechanical system) and/or is equipment (solid system) in the air. From ball games like golf, baseball, soccer, football and tennis to athletics, alpine skiing, cross-country skiing, ski jumping, cycling, motor sport and many others, the application of some basic principles of aerodynamic can make the difference between winners and losers.

If the general shape of the athlete/equipment system in terms of postural strategies and equipment customization is not optimized, it can either be made to deviate from its initial path, resulting in wrong trajectories and/or loss of speed and leading to failure in terms of performance. Coaches should thus be able to assess the aerodynamic efficiency of the motor task performed by the athlete with accuracy and in almost real time. Indeed, quick answers and relevant information can help the athlete to focus on specific aspects of his technical behaviour to improve his performance. So far for this purpose, two solutions are available i.e. dedicated wind tunnel testing or implementation of aerodynamic force models during the athlete training sessions. According to the complexity of sport performance and the necessity of almost real time answers for stakeholders, issue concerning the relevance of aerodynamic force modelling versus controlled experiments in wind tunnel must be discussed. In particular when searching to optimize athletes' performances, what are the advantages to develop and implement aerodynamic models comparing to controlled experiments in wind tunnel and for which purpose?

After a short description in section 2 of the aerodynamic principles commonly applied in sport to help optimize performance, the current chapter will document in section 3 both approaches (wind tunnel testing and aerodynamic force modelling) to assess the aerodynamics properties of a particular mechanical system: the athlete with or without his equipment. It will among others present a review of particular wind tunnel setting and modelling methods dedicated to specific sports such as cycling and skiing as well as shows in section 4, how appropriate applications of them can lead to an increase of athletes' performances.

2. AERODYNAMIC PRINCIPLES APPLIED TO HELP OPTIMIZE PERFORMANCE IN SPORT

2.1. The Performance In Sport

Athletic performance is a part of a complex frame and depends on multiple factors (Weineck, 1997). For sports such those involving running, cycling, speed skating, skiing ... where the result depends on the time required to propel the athlete's body and/or his equipment on a given distance, the performance is largely conditioned by the athlete technical skills. Success then is the outcome of a simple principle i.e. the winner is the athlete best able to reduce resistances that must be overcome and best able to sustain an efficient power output to overcome those resistances.

In most of the aforementioned sports, those resistances are mainly the outcome of the combination of the contact force and the aerodynamic force acting on the athlete (Fig. 1.) The goal in order to optimise the performance consists to reduce both of them as much as possible.

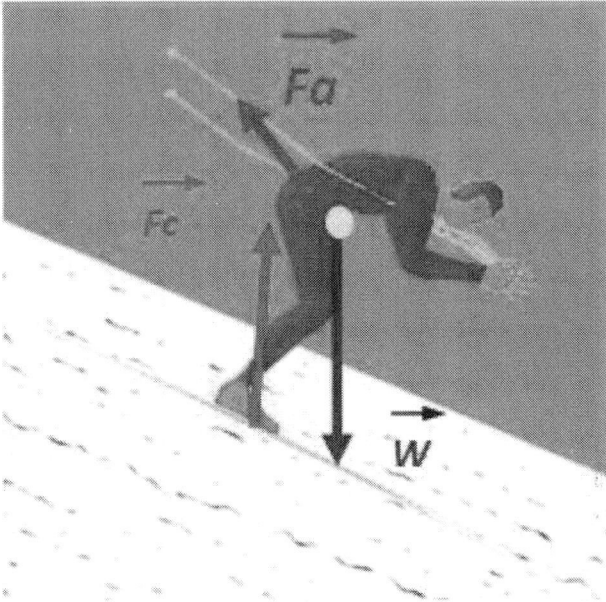

Figure 1. Force acting on a downhill skier. With W̶W̶ the weight of the skier, FeFethe ski-snow contact force and FaFathe aerodynamic force.

However, whether cycling, speed skating, skiing, given optimal physical capabilities, it has been shown that the main parameters that can decreased the race time considerably is the aerodynamic behaviour of the athlete and/or his equipment. Indeed, in cycling, the aerodynamic resistance is shown to be the primary force impeding the forward motion of the cyclist on a flat track (Kyle et al., 1973; Di Prampero et al., 1979). At an average speed close to 14 ms⁻¹, the aerodynamic resistance represents nearly 90% of the total power developed by the cyclist (Belluye & Cid, 2001). The statement is the same in downhill skiing. The aerodynamic resistance is the parameter that has the greatest negative effect on the speed of the skier. For a skier initially running with a speed of 25 ms⁻¹, the transition from a crouch posture to a deployed posture can induce in 2 seconds (1.8% of the total run) almost a decrease of 12% of the skier speed whereas in the same condition, the ski-snow contact force only lead to a decrease of 2.2% (Barelle, 2003).

It is thus obvious that in such sports where a maximal speed of the system athletes/equipment is needed in order to reduce as much as possible the racing time, an optimisation of the system aerodynamic properties is crucial compare to the optimization of its contact properties.

2.2. Fundamentals Of Aerodynamic

Aerodynamics in sport is basically the pressure interaction between a mechanic system (athlete and/or his equipment) and the surrounding air. The system in fact moves in still or unsteady air (Fig.2.).

Figure 2. A downhill skier passing over a bump (photo: Sport.fr).

By integrating the steady and static pressure field over the system, the resulting aerodynamic force acting on this system can be obtained (Nrstrud, 2008). This force is generally divided into two components, i.e. the drag force $\overset{o}{D}$ and the lift force $\overset{o}{L}$ (Fig.3.).

The drag $\overset{o}{D}$ is defined as the projection of the aerodynamic force along the direction of the relative wind. This means that if the relative wind is aligned with the athlete/equipment system, the drag coincide with the aerodynamic force opposite to the system motion. $\overset{o}{D}$ depends on three main parameters: (i) the couple athlete/equipment frontal surface area (defined as the surface area of the couple athlete/equipment projected into the plane perpendicular to the direction of motion), (ii) the drag coefficient depending on the shape and the surface quality of the system and (iii) the athlete speed. The drag is thus expressed using the following equation (1).

Figure 3. Aerodynamic force applied on a skier and its two components: $\overset{o}{D}$ the drag (axial component) and ⌐ the lift (normal component). V represents the speed of the skier.

$$D = \tfrac{1}{2} \circ \prec \circ A \circ C_D \circ V^2 \tag{1}$$

Where D denotes the drag (N), ρ is the air density (kgm^{-3}), A is the projected frontal area of the couple athlete/equipment (m^2), CD is the drag coefficient and V is the air flow velocity (ms^{-1}) equivalent to the athlete speed.

The drag is essentially proportional to the square of the velocity and its importance grows more and more as the speed increases. If speed is doubled, the drag increases by four-fold. The drag coefficient C_D is dimensionless and depends on the Reynolds number (ratio ofinertial forcesand forces due to the viscosity of air) and the speed of the airflow. If C_D varies for law speed values (Spring et al., 1988), in most of the sports considered in this chapter, it can be considered as constant (Di Prampero et al., 1979 ; Tavernier et al., 1994). In fact, the athletes never reach the critical speed which cause the fall in C_D due to the change from laminar to turbulent regime. So at a steady and relatively high speed, variations of drag are mainly induced by variations of the projected frontal area of the couple athlete/equipment, thus by posture variations

(Watanabe & Ohtsuki, 1977; 1978). The figure 4 shows in which proportion the A.C$_D$ factor of a downhill skier varies with changes in posture.

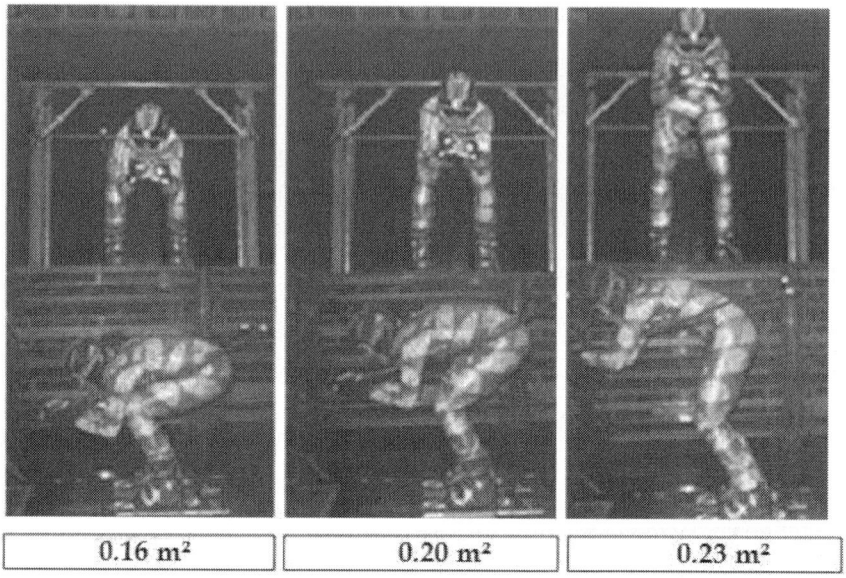

| 0.16 m² | 0.20 m² | 0.23 m² |

Figure 4. Variation of the A.C$_D$ factor of a downhill skier according to posture variations (Wind tunnel of IAT, France).

The lift L is the component of the aerodynamic force that overcomes gravity. It is acting normal to the drag component. As the drag, it depends also on three main parameters: (i) the couple athlete/equipment frontal surface area (defined as the surface area of the couple athlete/equipment projected into the plane perpendicular to the direction of motion), (ii) the lift coefficient depending on the shape and the surface quality of the system and (iii) the athlete speed. The lift is thus expressed using the following equation (2)

$$L = 1/2 \circ \rho \circ A \circ C_L \circ V^2 \qquad (2)$$

Where L denotes the lift (N), ρ is the air density (kgm⁻³), A is the projected frontal area of the couple athlete/equipment (m²), CL is the lift coefficient and V is the air flow velocity (ms⁻¹) equivalent to the athlete speed.

Bernoulli's lawexplainsthephenomenononofliftfrompressure differences between the lower and upper surfacesof theprofileof a mechanical system (Fig. 5).

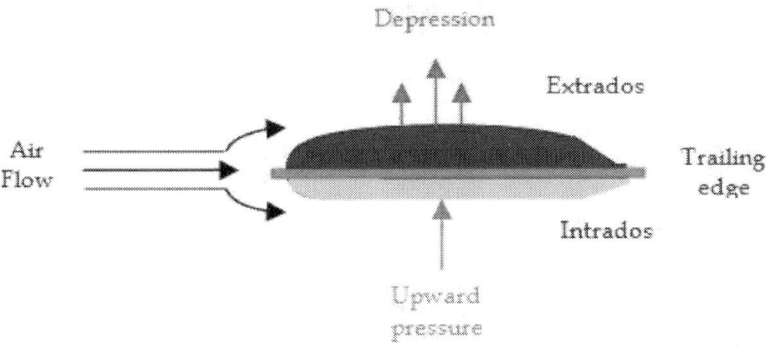

Figure 5. The lift effect according to Bernoulli's law.

The distance travelled by the air flow is more important above the extrados than below the intrados. To avoid creating a vacuum of air at the trailing edge, the air flow following the extrados must move faster than the one following the intrados. An upward pressure is thus formed on the intrados and a depression appears on the extrados, thereby creating a phenomenon of lift. The shape of the mechanical system and its surface quality have thus, an effect on the lift intensity. However in the same manner as the drag coefficient C_D, the lift coefficient can be considered constant for the ranges of speed practiced during the aforementioned sports. Variations of the surface opposing the airflow induced by variations of the angle between the system chord line and the longitudinal axis (Fig.6.) namely the angle of incidence (i), impact the variability of the lift (Springings & Koehler, 1990). For an angle of incidence greater than 0 °, the lift will tend to increase while for an angle of incidence lower than 0 °, a phenomenon of "negative lift" will appear (downforce).

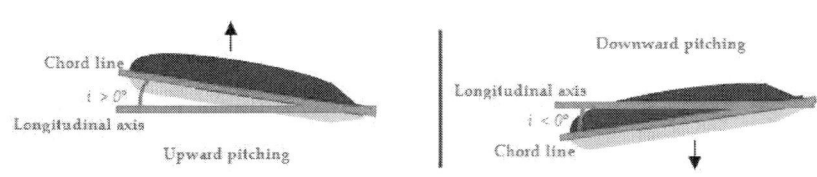

Figure 6. Profile of an object according to its angle of incidence. i correspond to the angle of incidence.

In the aforementioned sports (running, cycling, skiing, skating), the equipment surface is rather small with respect to the athlete surface and therefore the main part of the aerodynamic force acts on the athlete who can be regarded as bluff body (non streamed line body). The bluffness leads to the fact that the aerodynamic resistance is mainly pressure drag instead of friction drag and thus, on a general point of view, it's more important to reduce the frontal area than to reduce the wet area. Then as lift is generally not required, it's better to keep it as small as possible in order to avoid the production of induced drag. However, in particular sport like ski jumping, it is obvious that the flight length is sensitive both to lift and drag. Small changes in the lift and or drag can have important effect for the jump quality and the skier must find the right compromise between an angle of incidence that will lead to an increase of the lift but not to an increase of the drag. The athlete must thus produce an angular momentum forwards in order to obtain an advantageous angle of incidence as soon as possible after leaving the ramp (Fig.7.). If the forward angular momentum is too low, the flight posture will induce a high drag thus a law speed and a low lift, resulting in a small jump. Too much forward angular momentum on the other hand can increase the tumbling risk.

Figure 7. A ski jumper during the flight phase just after leaving the ramp (photo: Photo by Jed Jacobsohn/Getty Images North America).

2.3. Reducing The Aerodynamic Force To Optimize The Performance

Reducing the air resistance in sport events typically involved improving the geometry of the athlete/equipment system. Optimisation of the athlete postures as well as the features of his equipment is generally required since they have a pronounced impact on the intensity of the aerodynamic force.

Firstly, by proper movement of the body segments (upper limbs, trunk, lower limbs) in order to minimize the frontal surface area exposed to the air flow, the posture can become more efficient aerodynamically. For example, in time trial cycling, it is now well known that four postural parameters are of primary importance in order to reduce the drag resistance i.e. the inclination of the trunk, the gap between the two elbows, the forearms inclination with respect to the horizontal plan, the gap between both knees and the bicycle frame (McLean et al., 1994). The back must be parallel to the ground, the elbow closed up, the forearms tilted between 5° and 20° with respect to the horizontal and the knees closed up to the frame (Fig.8.).Such a posture (time trial posture) can lead to average reduction of the drag resistance of 14,95% compared to a classical "road posture" (37.8±0.5 N vs. 44.5±0.7 N; $p<0.05$)and that merely because of significantly lower frontal area (0.342±0.007 m2 vs. 0.398±0.006 m2; $p<0.05$) (Chabroux et al., 2008).

Figure 8. An optimal aerodynamic posture in time trial cycling.

In downhill skiing, the principle is the same. The intensity of the aerodynamic resistance is even lower that the skier adopts a compact crouched posture for which the back is round and horizontal, the shoulders are convex and the upper limbs do not cross the outer contour of the skier and especially do not obstruct the bridge created by the legs.

Figure 9. An optimal aerodynamic posture in downhill skiing on the left compare to a posture a little bit more open on the right (Wind tunnel of IAT, France).

For an initial skier speed of 25ms⁻¹, such a crouched posture can lead to a gain of 0,04 second after a straight run of 100 meters thus to a victory compared to a posture a little bit more open (Barelle, 2003).

Secondly suitable aerodynamic customisation of the equipment can also strongly reduce the negative effect of the aerodynamic resistance. Indeed as example, in cycling, the comparison between time trial helmet and normal road helmet shows a drag resistance improvement that can range from 2,4 % to 4 % according to the inclination of the head (Chabroux et al., 2008).

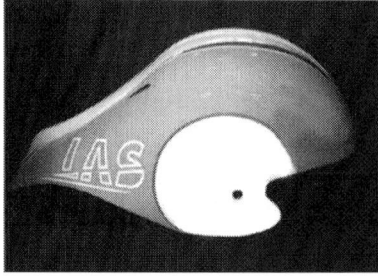 

Figure 10. Two cycling helmets, one aerodynamically optimised for time trial event (left) and the other a simple road helmet (right).

It is worth noting that an efficient optimisation of the aerodynamic properties of the athlete/equipment system must take into consideration precisely the interaction between the posture features and the equipment features. The aerodynamic quality of the equipment is totally dependent of the geometry characteristics of the athlete during the sport activity. An efficient optimization cannot be done without taking this point into consideration. In particular in time trial cycling, the interaction between the global posture of the cyclist and the helmet inclination given by the inclination of the head is significant from an aerodynamic point of view. The drag resistance connected with usual inclination of the head (Fig.11) is lower (37.2 ± 0.6 N) than the one related to the low slope of the head (37.8 ± 0.5 N), which is itself significantly lower than the one generated by a high slope of the head (38.5 ± 0.6 N). In fact according to the helmet shape, the inclination of the head can have different impact on the projected frontal area of the couple helmet /athlete head thus on the aerodynamic drag.

Hence, it is also important for coaches and athletes to optimize postures in a way that it will not affect the athlete physical power to counteract the resistance. In most of the sport and for aerodynamic purposes, athletes are asked to adopt a tightly crouched posture to reduce their frontal areas exposed to the air stream but if it is not well done, it can also have bad biomechanical and physiological consequences for the athlete performance such as a decrease of physiological qualities. Everything is a compromise. In ice skating for example, although a tightly crouched posture reduces leg power, it reduces air drag to an even greater extent and thus produces higher skating velocities.

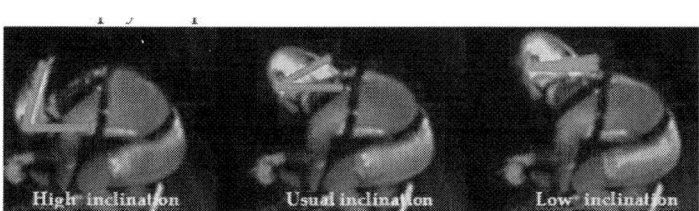

Figure 11. Inclination of the head in time trial and corresponding inclination of the helmet (Wind tunnel of Marseille, France).

3. METHODS FOR ASSESSING THE AERODYNAMIC FORCE APPLIED ON AN ATHLETE WITH OR WITHOUT HIS EQUIPMENT

To assess the aerodynamic performance of an athlete and/or his equipment, two methods are available, i.e. either to perform wind tunnel testing to single out only one specific determinant of the performance in this case aerodynamic properties of the athlete or/and his equipment, or to develop and implement aerodynamic force models that can for example be apply in a real training or competitive conditions which mystifies the role of other factors such as for instance mental factors. The real question here, concern the relevance of the inferences drawn from the results obtain with this two methods according to the fact that the performance in sport is the outcome of the efficient interaction of multiple factors at the right time. Indeed, "a fact observed in particular circumstances can only be the result of particular circumstances. Confirming the general character of such a particular observation, it is taking a risk of committing a misjudgement." (Lesieur, 1996). Both approaches are further detailed below as well as their relevance according to the performance goal pursue by the principles stakeholders i.e. athletes and coaches.

3.1. Wind Tunnel Testing

Wind tunnel tests consist in a huge apparatus used to determine the complex interactions between a velocity-controlled stream of air and the forces exerted on the athlete and his equipment.The tunnel must be over sized compare to the athlete to be assessed in order to avoid side effects that may disturb the measurement of the aerodynamic force. The athlete with or without his equipment is fasten on a measured platform (6 components balance) in the middle of the test section. The athlete is thus stationary in the flow field and the air stream velocity around him generally corresponds to the ones observed during the sport practice (e.g. 14ms^{-1} in time trial cycling, 25 ms^{-1} and more in alpine skiing.). The aerodynamic balance enables to measure the smallest aerodynamic force imposed on the athlete/equipment system in particular its axial (drag) and normal (lift) components (Fig.12).

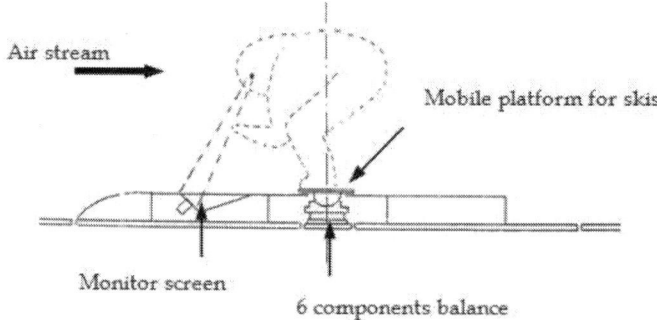

Figure 12. Diagram of a data acquisition system for the assessment of the aerodynamic properties of a downhill skier (Wind tunnel of IAT, France).

For a better understanding, the path of the air stream around the system can be made visible by generating smoke streams (Fig.13).

Figure 13. Smoke stream around a time trial cyclist and his equipment (Wind tunnel of Marseille, France).

A tomography gate can also be installed in the wind tunnel behind the athlete to explore the air flow wake behind him (Fig.14).

The figures below shows different wind tunnel settings that have been used for the measurement of the aerodynamic force applied on downhill skiers and time trial cyclists.

In alpine skiing, most of the time, the skier is in contact with the snow and only an accurate assessment of the drag applied on him is necessary. However in particular conditions and especially when he passes over a bump (Fig.2), it is interesting to quantify the lift applied on him. It has to be the smallest as possible since the skier as to be as soon as possible in contact with the snow to manage his trajectory. The length of the jump must be very short according to the initial and following conditions and the goal for the skier is to adopt in the air a posture that will generated the smallest lift. For both purposes i.e. measuring accurately the drag and the lift, two wind tunnel setting must be considered (Barelle, 2003;2004).

On Fig.15, the goal is only to measure the aerodynamic drag applied on a skier adopting a crouched posture. The measuring device is the one of the Fig.12. The skier is fastening in the middle of a wind tunnel (rectangular section, 5 meters wide by 3 meters in height and 10 meters length) on a 6 components balance that enables ones to have access to multiple variables, among other the aerodynamic drag. Wind-less balance signals acquisition (during which the skier has to keep the crouched posture) are generally performed before each aerodynamic measurement trial, in order to correct the measurements for zero drift and mass tares. After the zeros acquisition, the wind tunnel is started and when the required speed of the air flow is reached, the athlete can optimized is posture according to the strategy build with his coach. A mobile platform allowed him to adjust the posture of his legs whenever he wants according to the information he can read on the monitor screen.

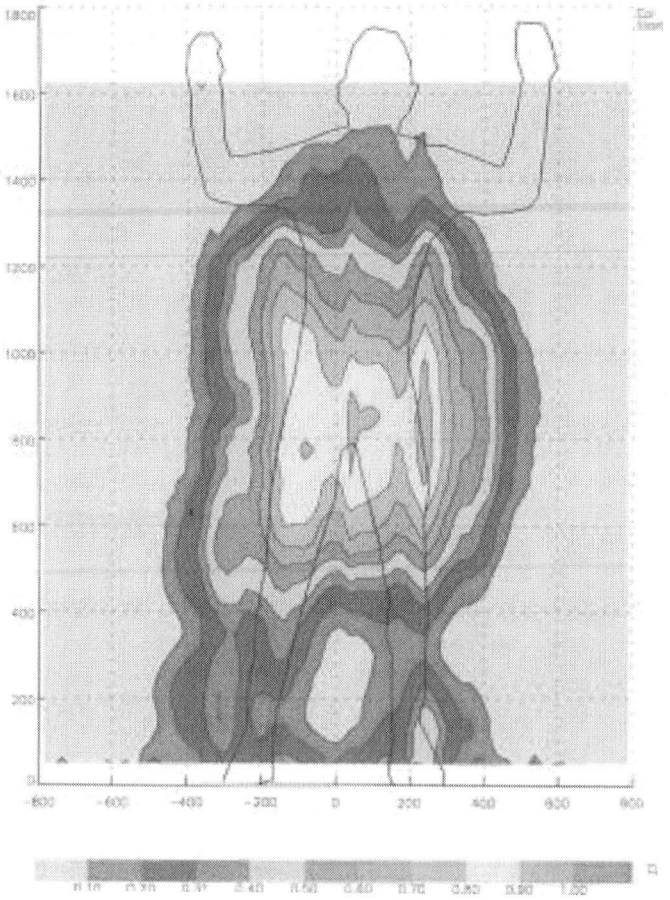

Figure 14. Mapping of the air flow behind a cross country skier (Wind tunnel of IAT, France). The more colours are warm, the more the aerodynamic resistance is important.

Figure 15. Measuring device for the assessment of the drag applied on a downhill skier (Wind tunnel of IAT, France).

If the skis have not a great impact on the variability of the drag intensity, their contribution to the variability of the lift has to be taken into account. It is therefore necessary to position the skis outside the boundary layer which is near the ground. Although it is relatively thin, the velocity of the airflow in this area varies significantly and disturbs the measurement of the lift. Sections of boat masts (Fig.16)located under each skis have thus allowed to overcome this problem and allowed to remove the skis from this thin layer where the air stream can transit from a laminar to turbulent conditions.

In time trial cycling, in order to determine the drag force of the system bicycle /cyclist, a cycletrainer is fastened on a drag-measurement platform mounted in the middle of the test- section of a wind tunnel which dimensions (octagonal section with inside circle of 3 meters in diameter and 6 meters length) allowed to avoid walls boundary layer effects that can interfering measurements (Fig.17). This platform is equipped with ball-bearing slides in the direction of the wind tunnel as well as a dynamometer measuring the drag force. As for assessing the aerodynamic properties of a skier, the general procedure for a cyclist is the same. A preliminary measurement without wind is performed in order to correct the measurements for zero drift and mass tares. Then a second measurement with wind but without the athlete allowed obtaining the drag force of solely the platform equipped with the cycletrainer. Finally, the drag force of the couple bicycle/cyclist can be measured while the cyclist adjusted his posture with a wind speed similar to that found in race conditions (around 14 ms^{-1}).

If such a measurement tools provides accurate recording of the aerodynamic force apply on the athlete, it has the disadvantages of not being able to be used anytime it is needed. Specific and dedicated wind tunnel program has to be perform and sometimes far away from the athletes current concerns. Moreover, the usual environmental conditions of the sport practice are requirements that cannot be taken into account in a wind tunnel setting.

Figure 16. Measuring device for the assessment of the lift applied on a downhill skier (Wind tunnel of IAT, France).

Figure 17. Measuring device for the assessment of the drag applied on a time trial cyclist.

3.2. Modelling Methods

For numerical models, the method consists in computing correlation between postural parameters observe during the practice as well as equipment characteristics when or if needed and the value of the aerodynamic force. It requires most of the time and previously wind tunnel data of the aerodynamic characteristics of the athlete according to various postures and if necessary within a wide range of orientations relative to the air flow (Fig.18). Indeed, the functions are generally determined with athletes or model of athletes positioned in a wind tunnel in accordance with postures observed during competition in the field.

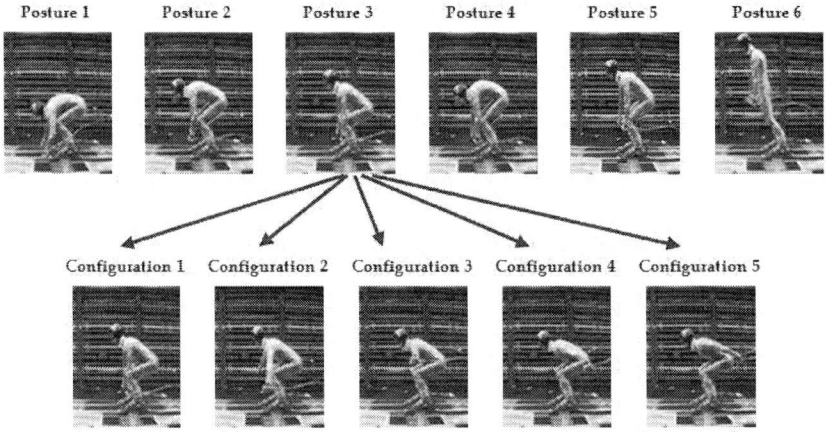

Figure 18. Postures assed in wind tunnel prior the development of a model of the aerodynamic lift applied on a downhill skier when passing over a bump. These postures correspond to postures observed in real conditions (Barelle, 2003).

The results of such models can then serve for example as input for simulations based on the Newton laws to estimate variations in time, loss in speed performance induced by different postural strategies as well as equipment interactions. When dedicated simulators integrating such models already exist, an almost real time feedback can be provided to the stakeholder on the aerodynamic properties of the athletes' posture. This can be a cost effective solution since it needs few human and material resources and it can be performed anytime it is needed during normal training sessions.

Examples of the development approach of some models for the evaluation of the aerodynamic performance in running, skiing, cycling are presented and discussed below.

3.2.1. Modelling Of The Aerodynamic Force In Running

Shanebrook & Jaszczak (1976) have developed a model for the determination of the drag force on a runner. They have considered the human body as a multi-jointed mechanical system composed of various segment and showed that the drag assessment applied on an athlete could be realized by considering the athlete's body as a set of cylinders. Their model is thus composed of a series of conjugated circular cylinders, to simulate the trunk and the lower and upper limbs, as well as a sphere to simulate the head. Projected surface area was measured for each segments (head, neck, trunk, arm, forearm, tight, shank) of the body of three runners representing respectively, adult American males in the 2.5, 50 and 97.5 percentiles of the population. Then the drag coefficient of cylinders and sphere representing these segments has been measured in a wind tunnel. The results for the 50 percentiles are proposed in the table here after (Table 1.).

If such a model has the merit to enable one to reach the drag coefficient of the body segments of a runner, it doesn't consider the athlete body has a whole as well as the succession of body segments orientations that can generate different projected surface area and thus variation of the air resistance throughout the global motion of the runner.

Table 1. Models to determine the drag coefficient of the body part of a runner according to their projected surface area according to Shanebrook & Jaszczak (1976).

	Cylinders	A (in²)	C_D
	1	64.5	1.2
	2	67.7	1.2
	3	67.7	1.2
	4	312	1.1
	5	78.1	1.2
	6	43.2	1.2
	7	11	1.2
	Sphere	48.3	0.43

Moreover the adaptation of such model to different runners or to different kind of sportsmen during their practice is time consuming and not in accordance with the stakeholders (coaches, athletes) requirement of a quick assessment of the aerodynamic performance of an athlete.

3.2.2. Modelling Of The Aerodynamic Force In Skiing

The aerodynamic resistance in alpine skiing has been largely investigated, leading to different approaches to model the aerodynamic force. Luethi & Denoth (1987) have used experimental data obtained in a wind tunnel in their approach of the aerodynamic resistance applied on a skier. They have attempted to assess the influence of aerodynamic and anthropometric speed skier. By combining the three variables most influencing the speed of the skier i.e. his weight, is projected surface area (reflecting its morphological characteristics), and the drag coefficient C_D, they established a numerical code (ACN:

Anthropometric Digital Code) representing the aerodynamic characteristics of skiers. The model is written as follow (3):

$$ACN = \left(\frac{mg}{A.C_D}\right)^{1/2} \tag{3}$$

Where m is the skier mass, A is the projected frontal area, CD is the drag coefficient.

If the factors mg and CD (invariable for skiers dressed with the same race clothes) are easily accessible, this model set the problem of assessing the projected frontal area of the skier in real condition. The observer (coaches) because of its placement on the side of the track can hardly have a front view of the athlete in action and even if he had it, it would not allow him to determine directly and easily the A. The model of Springings et al. (1990) for the drag and lift lead to the same problem. For this purpose, Besi et al. (1996) have developed a an images processing software to determine A but the processing time is once again too important for field application.

Spring et al. (1988) uses the conservation of energy principle in order to model the term $A.CD$ (4).

$$A.C_D = \frac{m\left(V_D^2 - V_F^2\right) - 2k.V.\,mg\,.d}{V^2.\!\div.d} \tag{4}$$

Where m is the skier mass, A is the projected frontal area, CD is the drag coefficient, VD is the initial speed of the skier, VF is the final speed of the skier, V is the mean speed of the skier, k the snow friction coefficient and the air density, d the distance travelled by the skier.

While this model takes into account as input data, field variables (speed of the skier, travelled distance), it does not incorporate the influence of postures variations. Once again the results obtained from this model can only be an approximation for use in real conditions since it cannot explain with accuracy the performance variations induced by changes in posture.

The modelling of the aerodynamic force as it is described above is not relevant and efficient for rapid application in real conditions. If in straight running, skiers can easily maintained an optimal crouched posture, in technical sections (turns, bumps, jumps), they must manage their gestures to ensure an optimal control of their trajectory, while minimizing the aerodynamic effects. To be relevant for such real conditions applications, posture variations must be taken into account in the modelling and thus whatever the considered sport.

3.2.3. Modelling of the Aerodynamic Force In Cycling

As cyclists' performances depend mainly on their ability to get into the most suited posture in order to expose the smallest area to the air flow action, the knowledge of their projected frontal area can be useful in order to estimate their aerodynamic qualities. By the way, several authors have either reported values of A or developed specific equations to estimate the projected frontal area (Gross et al., 1983;Neumann, 1992;Capelli et al., 1993; De Groot et al., 1995; Padilla et al., 2000;Heil, 2001). However, this has been generally done only for riders of similar size and adopting the same posture on a standard bicycle. Such estimations have then shown large divergences and methodological differences may have widely contributed to such variability. Thus to be useful, models mustn't be developed as black boxes but by indicating accurately why they have been develop for and in which condition they can be used, by being transparent on the variables that have served to its construction and the results accuracy it can provided.

For example, Barelle et al. (2010) have developed a model estimating accurately A as a function of anthropometric properties, postural variations of the cyclist and the helmet characteristics. From experiments carried out in a wind tunnel test-section, drag force measurements, 3D motion analysis and frontal view of the cyclists were performed. Computerized planimetry measurements of Awere then matched with factors related to the cyclist posture and the helmet inclination and length. A Principal Component Analysis has been performed using the set of data obtained during the experiment. It has shown that A can be fully represented by a rate of the cyclist body height, his body mass, as well as the inclination and length of his helmet. All the above mentioned factors have been thus taken into account in the modelling (5).

$$A = 0.045 \times h^{1.15} \times m_b^{0.2794} + \left[0.329 \times (L \times \sin \sin \Box_1)^2 - 0.137 \times \left(L \times \sin \sin \Box_1 \right) \right]$$

(5)

where h is the height of the cyclist, m_b the body mass of the cyclist, L the length of the helmet, and $_{a1}$the inclination of the head.

The prediction accuracy was then determined by comparisons between planimetry measurements and A values estimated using the model. Within the ranges of h, mb, L and αl involved in the experiment, results have shown that the accuracy of the model is 3%. Within the objective to be easy to use, this accuracy can be considered sufficientenough to show the impact of postural and equipment changes on the value of the frontal area of cyclists. This model is explicit and it has been developed to take into consideration variation of posture i.e. inclination of the head. It can easily be applied to a variety of cyclists with different anthropometric characteristics since the height and body mass are input data. Moreover it can also considered the shape characteristic of the helmet including (L) its interaction with the inclination of the head (αl). Finally its conditions of use are specified since its accuracy can only be guaranteed for input data that are within the ranges of h, mb, L and αl involved in the experiment. It can thus provide pertinent indications useful for both coaches and cyclists.

3.3. On The Relevance Of Aerodynamic Force Modelling Versus Wind Tunnel Testing

Individual and accurate optimization of the aerodynamic properties of athletes on very details modifications by means of wind tunnel measurements is essential for high performance. However, such comprehensive experiments in large scale wind tunnels lead to excessive measurement time and costs and require the disposability of athletes over unreasonably long periods. Even if accurate, wind tunnel tests have the disadvantage of not being able to be used anytime it is needed as it is required for high level sport. Moreover, the usual environmental conditions of the sport practice that can widely influence the performance are requirements that cannot be taken into account in a wind tunnel setting.

Instead, the computer modelling approach if well oriented allows studying the impact of all variables, parameters and initial conditions which determine the sport performance. In terms of aerodynamic, models implemented in the years 1980 and 1990 (Shanebrook, 1976; Watanabe & Ohtsuki,1978; Luethi et al., 1987; Springings et al., 1990...), do not report the low dispersion of athletic performance neither because of the technical means available for their implementation nor because they were not designed for this purpose.

Several authors have tried to formalize the different steps to develop useful model (Vaughan, 1984;Legay, 1997) but this process is not as linear as it seems. The first stage involves identifying the system under study. This is a situation analysis which will determine and describe the framework within which will take place all the work ahead. When the frame is set, it is about to implement procedures to collect data relating to the objective pursued. The choice of tools for collecting and processing experimental data must be consistent with the model and the desired accuracy. Wind tunnel testing can thus in this case be useful if it takes into consideration postures observed during training and racing, athlete/equipment interactions, boundary conditions. Then to build the model, dependencies between different recorded variables are considered. These relationships are then translated in the form of equations giving the model structure. According Orkisz (1990), it must be hierarchical and give the possibility to adapt to all levels of complexity, depending on the nature of the results to be obtained. Such models have an important value in the quest for performance if their results are express in term of objective benchmarks (time, speed, trajectories...) that can extend the observation of the coaches.

They could have two exploitation level i.e. analytical or global since they enable stakeholders respectively to focus on a particular aspect of performance such as the specific influence of the aerodynamic resistance (analytical approach of the Newton's law) or on the interaction of factors determining the performance (global approach of the Newton's law)with the aerodynamic resistance among others (Barelle, 2003). When such models are used for simulation, they allow stakeholders to go further than the simple description. Beyond the fact that they can be used anytime it is needed, they have also predictive capacities and that, at a lower cost.

4. APPLICATION AND VALORISATION: TOWARDS AN OPTIMIZATION OF DOWNHILL SKIERS' PERFORMANCES WHEN PASSING OVER A BUMP

For each discipline in Alpine skiing (downhill, slalom, giant slalom...), the difference in performance among the top world skiers is lower than one percent. Taking into account this low variability, coaches are confronted with the problem of assessing the efficiency of different postural strategies. Numerical models may provide an adequate solution. The method consists in computing a correlation between skiers' kinematics and postural parameters observed during training and each of the forces involved in the motion's equation (Barelle, 2003, Barelle et al., 2004; Barelle et al.; 2006). For postural strategies such as pre-jump or op-traken in downhill, models of the projected frontal area for the lift (6)(Barelle, 2003) and for the drag (7) (Barelle et al., 2004) are calculated based on postural parameters (length and direction of skier's segments).

$$A_L = 0.1167 \sin(\vartheta) + 0.0258 \sin(\beta) + 0.0607 + 0.024 E_T ((\sin \sin(2.\varpi_3) - \cos(\varpi_4))$$

(6)

Where A_L is the projected frontal area, is the orientation of the trunk, is the orientation of the tight in the sagittal plan, 3 and 4 are the arms orientation respectively in the frontal and horizontal plan.

(7)

Where A_D is the projected frontal area, is the orientation of the trunk, β is the orientation of the tight, α is the orientation of the shank in the saggital plan, 1 and 2 are both arms orientations in the horizontal plan.

Ground reaction and skis-snow friction are computed according to skiers' postural kinematics (skier's amplitude variation and duration of spread movements). Skiers' weight is easy to obtain. Thus the external forces exerted on the skis-skier system (Fig.1) are known, the motion's equation can be solved and simulations performed (Fig.19). These can be used to estimate variations in time and loss in speed performance induced by different postural strategies.

Such simulations find an application in the field of training as they enable to assess the impact on performance of a given strategy compared with another (Barelle, 2003; Barelle et al., 2006). Simulation results can be presented in the form of animations, using DVD technology. Such tool enables trainers to show skiers very quickly the variability of performance induced by different postural strategies (Fig.20.).

Broken down in this form, the simulation becomes a way of learning transmission. The aerodynamic drag model(7) can be used directly, if the coach chooses to particularly focus his attention on the aerodynamic effects. Afirst level of use is then given to the model. Then the model can have a second level of use, if the coach wants to have a general view of the skier performance since it is also designed to be an integral part of the modeling of the postural strategies implemented by skiers when passing over a bump in downhill skiing (simulator, Fig.19.).

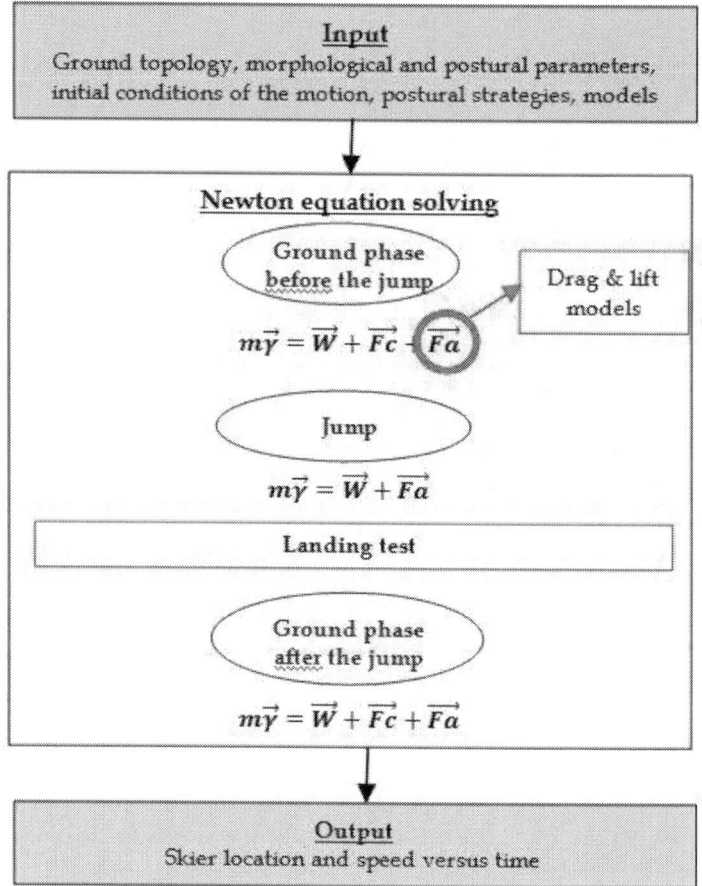

Figure 19. Structure overview of the simulator of the trajectory of the centre of mass of a skier according to his anthropometric characteristics and his postural strategy as well as the topology of the downhill slope.

5. ACKNOWLEDGEMENTS

Researches on downhill skiing are a compilation of several wind tunnel tests (Wind tunnel of IAT, France) conducted each years from 2000 to 2003 by the French Ski Federation in order to optimize the downhill posture of its athletes. The author wishes to thanks particularly all the coaches and skiers that have widely contribute to obtain such results.Researches on time trial cycling were performed in 2007 (Wind tunnel of Marseille, France) and supported by a grant between Bouygues Telecom, Time Sport International and the University of Mediterranean. The author wishes to thank all members of the cycling team for their active contribution to the wind tunnel testing campaigns.

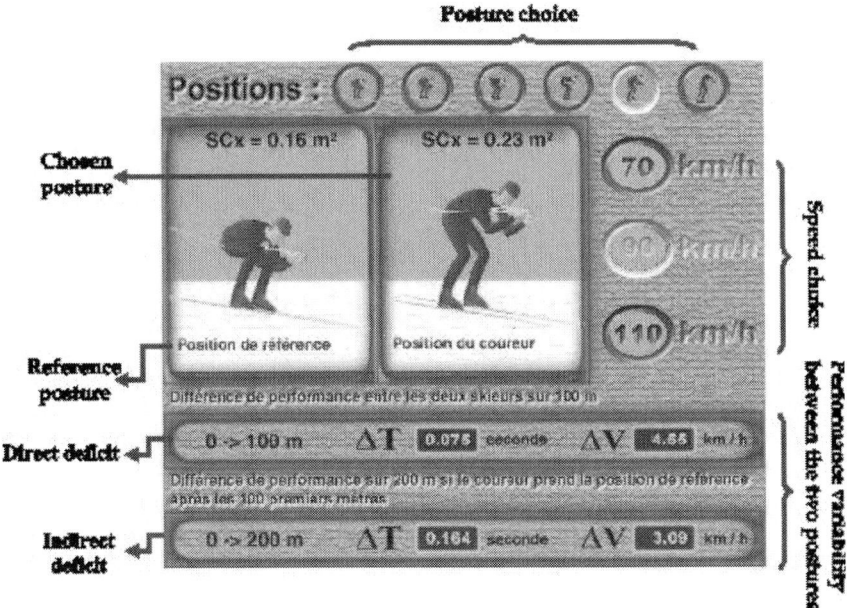

Figure 20. Overview of DVD application built for the downhill skiers of the French Ski Federation. The choice of a posture enables ones to see the aerodynamic drag impact on performance for three input speed. The choice of a particular input speed enables to see the aerodynamic drag impact according to six different postures usually observed during races. The direct performance variability in terms of time deficit and loss of speed between the reference posture and the chosen posture is given after 100 meters of straight running (Direct deficit). Then stakeholders can visualize the indirect deficit generate 100 meters further (200m) even if the skier adopt again an aerodynamic crouched posture (like the reference one) on the last 100 meters (Indirect deficit).

REFERENCES

1. C. Barelle, 2003 Modélisation dynamique du geste sportif à partir de paramètres posturaux. Application à l'entraînement en ski alpin. PhD Thesis, Claude Bernard University, Lyon, 99102 .
2. C. Barelle, A. Ruby, M. Tavernier, 2004 Experimental Model of the Aerodynamic Drag Coefficient in Alpine Skiing. Journal of Applied Biomechanics,20 167176 .
3. C. Barelle, A. Ruby, M. Tavernier, 2006 Kinematic analysis of the performance based on simulations of the postural strategies produced by the alpine skiers. Science et Motricité, 3 59 99111 .

4. C. Barelle, V. Chabroux, D. Favier, 2010 Modeling of the Time Trial cyclist projected frontal area incorporating anthropometric, postural and helmet characteristics, Sports engineering,12 4 199206 .

5. N. Belluye, M. Cid, 2001 Approche biomécanique du cyclisme moderne.Science et Sports, 16 7187 .

6. M. Besi, D. D. Vedova, L. M. Leonardi, 1996 Sections : un programma di analisi dell'immagine applicato allo sport. Scuola dello sport. 34 7277 .

7. V. Chabroux, C. Barelle, D. Favier, 2008 Aerodynamics of time trial bicycle helmets. The engineering of sport, 7 401410 .

8. C. Capelli, G. Rosa, F. Butti, G. Ferretti, A. Veicsteinas, P. E. Di Prampero, 1993 Energy cost and efficiency of riding aerodynamic bicycles. European Journal of Applied Physiology, 67 149165 .

9. G. De Groot, A. Sargeant, J. Geysel, 1995 Air friction and rolling resistance during cycling. Medecine and Science in sports and exercise, 10901095 .

10. P. E. Di Prampero, G. Cortili, P. Mognoni, F. Saibene, 1979 Equation of motion of a cyclist. Journal of applied physiology, 47 201206 .

11. A. C. Gross, C. R. Kyle, D. J. Malewicki, 1983 The aerodynamics of human-powered land vehicles. Scientific American, 249 126134 .

12. D. P. Heil, 2001 Body mass scaling of a projected frontal area in competitive cyclists. European Journal of Applied Physiology, 85 358366 .

13. C. R. Kyle, C. Crawford, D. Nadeau, 1973 Factors affecting the speed of bicycle.Engineering report 731 . California State University. Long Beach, California.

14. J. M. Legay, 1997 L'expérience et le modèle : un discours sur la méthode. Sciences en question. INRA éditions.

15. P. Lesieur, 1996 L'étude de cas : son intérêt et sa formalisation dans une démarche clinique de recherche. Colloque interface INSERM / FFP.

16. M. S. Luethi, J. Denoth, 1987 The influence of aerodynamic and anthropometric factors on speed in skiing. International journal of sport biomechanics, 4 345352 .

17. S. Padilla, I. Mujika, F. Angulo, J. J. Goiriena, 2000 Scientific approach to the 1h cycling world record : a case study. Journal of Applied Physiology, 89 15221527 .

18. B. D. Mc Lean, R. Danaher, L. Thompson, A. Forges, G. Coco, 1994 Aerodynamic characteristics of cycle wheels and racing cyclists. Journal of Biomechanics, 27 675

19. G. Neumann, 1992 Cycling. Endurance in sport.Edition R.J. Shepard and P.O. Astrand, London : blackwell, 582593 .

20. H. N(rstrud, 2008 Basic Aerodynamics. Sport aerodynamics. CISM International centre for mechanical sciences, 506 18 .

21. M. Orkisz, 1990 Traitement d'image pour l'analyse du mouvement humain. Cinesiologie, 29 133140 .

22. R. J. Shanebrook, R. D. Jaszczak, 1976 Aerodynamic drag analysis of runners. Medecine and science in sports, 8 1 4345 .

23. E. Spring, S. Savolainen, J. Erkkilä, T. Hämäläinen, P. Pihkala, 1988 Drag area of a cross country skier. International journal of sport biomechanics. 4 103113 .

24. E. J. Springings, J. A. Koehler, 1990 The choice between Bernoulli's or Newton's model in predicting lift. International Journal of sport biomechanics, 6 235245 .

25. M. Tavernier, P. Cosserat, E. Joumard, P. Bally, 1994 Influence des effets aérodynamiques et des appuis ski- neige sur la performance en ski alpin. Science et motricité, 21 2126 .

26. K. Watanabe, T. Ohtsuki, 1977 Postural changes and aerodynamic forces in alpine skiing. Ergonomics, 20 2 121131 .

27. K. Watanabe, T. Ohtsuki, 1978 The effect of posture on the running speed of skiing. Ergonomics, 21 12 987998 .

28. C. L. Vaughan, 1984 Computer simulation of human motion in sports biomechanics. Exercice and sport sciences reviews, 12 373416 .

29. J. Weineck, (1997) Manuel d'entraînement. Vigot, collection Sport + Enseignement. 4ème édit ion.

CHAPTER 4

Propulsion for Biological Inspired Micro-Air Vehicles (MAVs)

Jorge M. M. Barata, Fernando M. S. P. Neves, Pedro A. R. Manquinho, Telmo A. J. Silva

Aerospace Sciences Department, Universidade da Beira Interior, Covilhã, Portugal

ABSTRACT

Small Unmanned Aerial Vehicles have been receiving an increasingly interest in the last decades, fostered by the need of vehicles able to perform surveillance, communications relay links, ship decoys, and detection of biological, chemical, or nuclear materials. Smaller and handy vehicles Micro Air vehicles (MAVs) become even more challenging when DARPA launched in 1997 a pilot study into the design of portable (150 mm) flying vehicles to operate in D^3—dull, dirty and dangerous—environments. More recently DARPA launched a Nano Air Vehicle (NAV) program with the objective of developing and demonstrating small (<100 mm; <10 g) lightweight air vehicles with the potential to perform indoor and outdoor missions. The current investigation is focused on the mechanisms involved with natural locomotion (propulsion and lift should not be considered independently). Biological systems with interesting applications to MAVs are generally inspired on flying insects or birds; however, similarly to the aerodynamics of flight, powered swimming requires animals to overcome drag by producing thrust. Commonalities between natural flying and swimming are analyzed together with flow control issues as a purpose of improvement on biology-inspired or biomimetic concepts for Micro Air Vehicles implementation.

KEYWORDS

Biomimetics, Flapping Wings, Insect Flight, Adaptive Biology

1. INTRODUCTION

This paper is focused on the mechanisms involved with natural locomotion (thrust and/or lift). Commonalities between natural flying and swimming are analyzed together with flow control issues. The present study has been driven by the ability of living organisms to fit an ecological system in terms of their locomotion. Historically, it was envisaged that men would fly by flapping artificial wings like birds; their physiological and biomechanical flapping flight procedures have been explored by men since Giovanni Alphonso Borelli [1] . On the XIX Century, Étienne Jules Marey developed studies about the insects flapping flights and was the first to notice a complex horizontal 8 shape wing motion pattern on its trajectory during the flight. In 1874 Pettigrew Bell published a book [2] on which he drew attention to the fact that the birds while flying and during every cycle of wingbeat, run movements that could be represented with considerable accuracy with an 8-figure drawn vertically, while the insects run the same figure drawn horizontally displaced. In 1902, Pettigrew stated that in a way to confer on the insect's wings the multiplicity of movements which they require, they are supplied with double hinge or compound joints, which enable them to move not only in an upward, downward, forward, and backward direction, but also on several intermediate degrees of obliquity—meaning this that insects wings are actually acting as helices, or twisted levers, and elevating weights much greater than the area of the wings would seem to warrant [3] . Similarly, by studying the fish's species, men found that most of them swim with lateral body undulations running from head to tail, in motion that also remind a 8-shape figure configuration, when viewed from the top (see Figure 1).

2. BACKGROUND

Small Unmanned Aerial Vehicles (UAVs) have been receiving an increasingly interest in the last decades. This interest was fostered by the need of vehicles able to perform surveillance, communications relay links, ship decoys, and detection of biological, chemical, or nuclear materials [4] . Smaller and handy vehicles (Micro Air Vehicles or MAVs) become even more challenging when DARPA launched in 1997 a pilot study into the design of portable (150 mm) flying vehicles to operate in D³—dull, dirty and dangerous—environments [5] . More recently DARPA launched a Nano Air Vehicle (NAV) program with the objective of developing and demonstrating small (<100 mm) lightweight air vehicles (<10 g) with the potential to perform indoor and outdoor missions [6] . All requirements of low altitude, long flight duration at low speeds (up to 100 km/h), small wing spans and masses, together with demanding capabilities of takeoff, climb, loiter, hover, maneuver, cruise, stealth and gust response are further beyond today's fixed wing or rotorcraft vehicles. At the same time, MAVs fit in the general sizes, weights, and locomotion performance of natural flying or swimming animals [7] . Nevertheless, biomimetic engineered devices are still far from the living organisms and more research is needed [8] . There is a general agreement that an unsteady dynamics approach is required to capture the physical phenomena at this scale [9] . Additionally, propulsion and lift should not be considered independently. Flapping wing systems appeared in animals such as insects, bats, birds, and fishes, which are known to exhibit remarkable aerodynamic and propulsive efficiencies. So, there have been several experimental and numerical studies of the biomimetic propulsive flapping [10] [11] . Most of these studies addressed the role of kinematic parameters such as flapping frequency, amplitude and phase difference on thrust generation and propulsive efficiency. At the same time, the effect of airfoil configuration has been considered far less and the published work is not always in agreement. For example, the results [12] - [14] show that thick airfoils can improve plunging airfoil performance, whereas [15] -[17] suggest than thin airfoils perform better, and the inviscid analysis [16] concludes no influence of airfoil thickness on plunging airfoil propulsion. Some authors attribute the superior efficiency of natural systems of thrust generation and propulsive efficiency to wing flexibility and focused their research on flexible wings with chord and span flexibilities [18] [19] . Has been reported [20] that flapping wings induce three rotational accelerations: angular, centripetal and Coriolis in the air near to the wing's surface, which diffuse into the boundary layer of the wing. Their results suggest that swimming and flying animals could control de predictability of vortex-wake interactions, and the corresponding propulsive forces with their fins and wings. Researchers [21] investigated dimensionless numbers to study swimming and flight, and their findings were disappointing since it became clear that different points of view exist in the biomechanics field on how to best define and use. So, successful biology-inspired or biomimetic concepts will depend on the understanding of the natural mechanisms especially when they do not agree with the present engineering design principles.

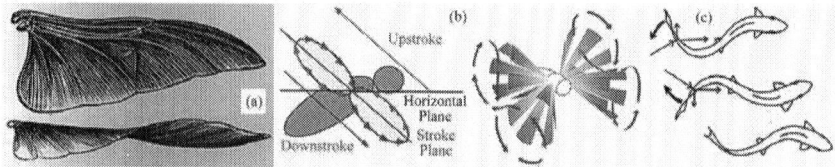

Figure 1. (a) Illustrative examples of the form and deformation of wings alluded to those of the beetle, bee, and fly—Petti- grew Bell [3] ; (b) lateral and isometric view of a generic insect wingstroke plane revealing a horizontal 8-shape figure drawn; (c) top view of fish's motion revealing an 8-shape figure drawn.

3. SOME CONSIDERATION RELATED TO FLUID MEDIA (WATER AND AIR)

Water and air are both regarded as fluid media, however, they are distinguished from each other: water is comparatively heavy ($\sim$1 ton/m^3) and incompressible; air, on the other hand, is comparatively light ($\sim$1.225 kg/m^3 at sea level, at 15°C) and incompressible below Mach number M $\sim$ 0.3 (the ratio of flow velocity/sound velocity must be greater than $\sim$0.3 for a change fluid density of >5%); for M > 0.3, significant compressibility occurs; all insect's fly in an incompressible air flow. When an animal swim through the water, the drag obtained is much greater when compared with the drag obtained from air similarly treated. Unsteady water flows are very common in nature, yet the swimming performance of fishes is typically evaluated by researchers, at constant, steady speeds at an appropriate facility. Similarly, most studies of insect flight are conducted in smooth flow or still air conditions. On both cases, it is still mostly unknown if unsteady water flows represent advantages and/or disadvantages to swimming fishes, and as well, if variable wind in nature affects flying insects as an advantage and/or disadvantage; however, in order to meet such peculiar requirements, all traveling organs of aquatic and flying animals (feet, fins, flippers, or wings) are not designed by nature as of rigid materials; instead, they are elastic materials.

4. SWIMMER ORGANISMS: LOCOMOTION ON FISH: TAIL AND FINS—CONTROL SURFACES

Most fish species swim with lateral body undulations running from head to tail, by exerting force against the surrounding water; these waves are slower than the waves of muscle activation; i.e., they contract muscles on either side of its body in order to generate flexion waves that travel the length of the body from head to tail. Fish's body is often fusiform, a streamlined body plan often found in fast-moving fish. They may also be filiform (eel-shaped) or vermiform (worm-shaped). Also, fish are often either laterally thin (compressed), or dorso-ventrally flat (depressed). Their muscle power is converted to thrust either directly by the bending body or almost exclusively by the tail, depending upon the body shape of the species and the swimming kinematics. Comparative scientists (physiologists and neurobiologists) have long been interested to realize how locomotion mechanisms used by aquatic organisms, propel themselves through water. The main external features of the fish are fins, composed of bony spines protruding from the body with skin covering them and joining them together; and as they are located at different body's places on the fish, they serve as well for different purposes, such as moving forward, turning, and keeping an upright position. Dorsal fins are located on the back: most fishes have one dorsal fin, but some fishes have two or three (as well as also could have finlets—small fins, generally between the dorsal and the caudal fins). The dorsal fins serve to protect the fish against rolling. The caudal fin (tail) is located at the end of the caudal peduncle and is used for propulsion. The tail fin can be: rounded at the end; truncated; forked; emarginated; or continuous. The anal fin is located on the ventral surface and is used to stabilize the fish while swimming. The pectoral fins are located on each side of the fish. A peculiar function of pectoral fins, highly developed in some fish, is the creation of the dynamic lifting force that assists sharks, in maintaining depth; they also enables the "flight" for flying fish and the "walking" in some anglerfish in the mud. The ventral fins assist the fish in going up or down through the water, turning sharply, and stopping quickly. Torsional angles changes on fish's fins can be produced by active control, via muscles force, or by passive control, via inertial hydrodynamic forces (Figure 2(a) and Figure 2(b)). Extensive studies have been made on the Dolphins [22] ; their fusiform and streamlined body shape, reduce the pressure component of the drag through maintenance of laminar flow; their maximum thickness (where transition to turbulent flow and boundary-layer separation is likely to develop) it is nearly at 45% of a body length from the beak, meaning that at least 45% of dolphins body may have laminar flow, due to a favorable pressure gradient up to the maximum thickness. The dolphins fineness ratio (FR = body length/maximum diameter) may range among the values $3.85 \leq FR \leq 5.55$ close to the optimum value of lowest drag of FR = 4.5 (e.g. [22]). Their propulsive movements are confined to the vertical plane in the posterior 1/3 of the body, with greatest amplitude at the caudal peduncle; the anterior body part acts as an inertial mass, minimizing energy loss from body oscillations. Dolphins could perform maximum speed up tom 9.3 ms^{-1}. Orcinus orca could perform maximum speed up to 15.5 ms^{-1} during 20 minutes. The pectoral flippers of the humpback whale (Megaptera novaeangliae) show leading edge tubercles [23] as illustrates Figure 2(c). Comparisons of wing sections with and without tubercles using CFD models, showed a 4.8% increase in lift; 10.9% reduction in induced drag; 17.6% increase in lift to drag ratio for wing section with tubercles at 10° angle of attack. Enhanced maneuverability by the addition of leading edge tubercles has potential application in the development of modern vehicles operating in air or water. Experimental flow visualization [24] compared with numerical simulations on both velocity and vorticity fields, revealed a good agreement; those results also revealed that fish can control body-generated vorticity, through body flexure and active manipulation by the caudal fin. Other researchers [25] , using DPIV, approach the question of why some fishes are able to swim faster than others, from a hydrodynamic perspective (Figure 2(d)); they investigate the structure and strength of the 3D wake to determine how hydrodynamic forces varies on Black surfperch (Embiotoca jacksoni) and Bluegill sunfish (Lepomis macrochirus). Both species (similar in size) swim at low speed using pectoral fins exclusively and both species at high speed, switch to pectoral-caudal fin locomotion; the surfperch can swim twice fast. It was found that surfperch presented a pair of linked vortex rings for all velocities, while the sunfish for low speed, presented only one vortex ring per fin and a pair of linked vortex rings with one ring only partially complete and attached to the body, at maximum speed. One of the most striking aspects of fish diversity is the presence of multiple locomotor control surfaces playing hydrodynamic roles during steady swimming and unsteady maneuvering locomotion, as well as the vortex wake interactions among all fins, a subject to be fully understood in the future. The skin of fast-swimming sharks (mako sharks) [26] is composed by a tooth-like scales (denticles) that generated vortexes on the front edge of the skin, i.e., eddies that essentially would help to pull the shark forward; this kind of skin composition is not found on slow-swimming sharks. Many researchers study this skin, by direct mimicking in its 3D shape or in a simplified grooved surface (riblets). Upon close examination of a dolphin's skin revealed micro dermal ridges that delay the transition to turbulent, by trap water molecules at the surface of the skin. Thus, the molecules of trapped water on the skin surface allows the animal to pass through the water more easily than if the same animal had a dry skin surface. The sailfish (that has V-shaped protrusions on skin) is known as fastest sea animal, reaching maximum speeds exceeding 110 km/h. His fin on the back which grows along the back can be spread and folded at will. Since sailfish is the fastest-swimming animal, researchers expected that sailfish's skin textures might produce skin-friction reduction; however, directly measures and numerical investigation showed increases or negligible reduction (~1%) on skin-friction. Yet, scientists advanced other explanations; the role of sailfish skin knowledge is to be confirmed.

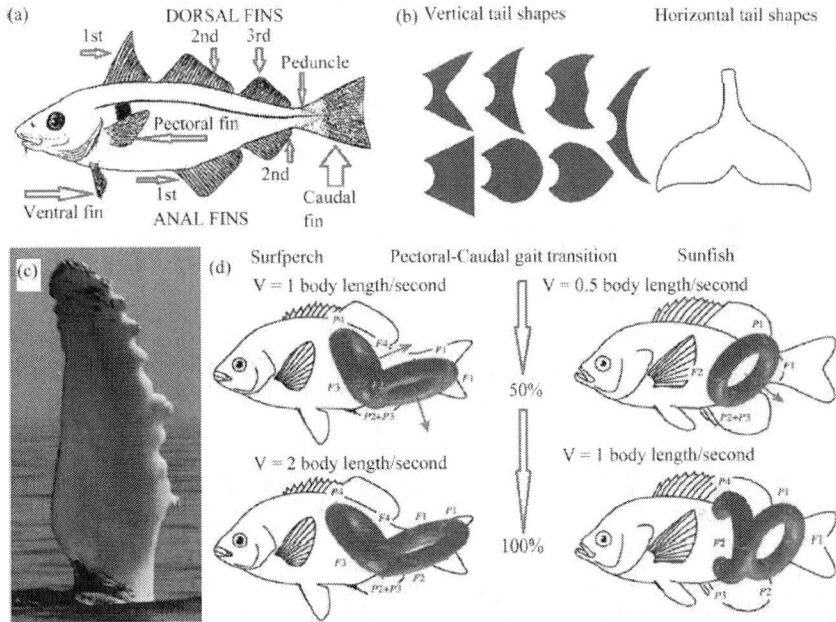

Figure 2. (a) Control surfaces on a generic fish; (b) Different types of fish fin tails known; (c) Photo of a pectoral flipper of the humpback whale (Megaptera novaeangliae) show leading edge tubercles [23] ; (d) Black surfperch (Embiotoca jacksoni) and Bluegill sunfish (Lepomis macrochirus), revealing vortex rings at different speeds [25] .

5. INSECT: A BIOLOGICAL FLIGHT MACHINE AND THEIR WING'S KINEMATICS

Every insect's wings when in motion are deformed by either the aerodynamic forces from the surrounding air flow, or by the inertial acceleration; the overall wing deformation is a combination of both and is in a continuously and constantly changing. The power product of the flight's muscles is transmitted to the wings which unlike an aircraft wings are neither streamlined nor smooth: the shape, corrugation and performance of the wings and the complex flapping motion during each stroke cycle will determine the ability of an animal to fulfil successfully every stunning maneuver. Each wingstroke cycle is typically divided into two translational phases: upstroke and downstroke; and two rotational phases: pronation and supination. In the forward-downstroke movement—main power stroke—the wing initiate the downward loop with a high angle of attack until the leading edge tilts downwards, where the wing momentarily becomes horizontal in the middle of the stroke, minimizing the angle of attack; stalling is prevented due to the fastest moving of the wing at this point. During the recovery stroke, when the wing moves upwards and backwards, the leading edge tips backwards. The wing is rotated again at the top of the recovery stroke, restoring the maximum angle of attack immediately before the next downstroke movement initiation. Every motion streamed to the wing lies in a composition of rotational, horizontal, vertical and torsional movements. Torsional angles changes on insect's wings can be produced by active control, via muscles force, or by passive control, via inertial aerodynamic forces. Insect's flight maneuverability is remarkable and by far, superior to every maneuverability of any man-made flying vehicle. Dickinson et al. [27] , stated that the aerodynamic performance of insects results from an interaction of three distinct, yet, interactive flight mechanisms: delayed stall, meaning that the wing sweep through the air with large angle of attack, during the translational portions of the stroke; rotational lift, meaning an augment in angle of attack at the end of the stroke, providing extra lift; and wake capture, meaning that the wing will capture some energy, left behind on the air by the previous wingstroke, providing extra power. Dragonfly wings possess great stability and high load-bearing capacity during flapping flight, gliding and hovering, despite the fact that their mass is less than 2% of the insect's total mass; such wings (forewings—front wings; and hindwings—rear wings) are composed by a thin cuticular membrane, supported by a vein system (venation) [28] . This venation structure consists on a net of veins (of different sections) that forms rectangular frames at the leading edge and hexagons or polygons with more than four sides at the trailing edge, allowing the requirements for different wing zones bearing different loads. Adding to this, their wings are highly corrugated (more corrugated at first 1/4 at the leading edge), which increases significantly the stiffness and strength of the wings and results in a lightweight structure with very good aerodynamic performance. The flow induced by the motion of insect's wings is highly unsteady and vortical. The aerodynamics between flapping and gliding flight differ substantially in two important ways: in a gliding wing, the air tends to remain attached and flowing smoothly over the surface of any airfoil; by contrast, the air over a flapping wing tends to become entrained in a swirling vortex bound to the upper surface of the wing—separated flow. And whereas the attached flow over a gliding wing look approximately similar from one moment to next, the separate flow over a flapping wing varies constantly—unsteady flow. Nowadays it is widely accepted that the insects make an extensive use of unsteady separated flow mechanisms in order to generate far greater aerodynamic forces that for them, would be impossible to achieve with steady, or quasi-steady, attached flow. In fact, insect's generate enough force to keep themselves in the air, because they flap their wings at a very high angle of attack that creates a structure at the leading edge of the wing, (a tornado-like structure) called leading edge vortex (LEV—see Figure 3). Large LEV's are formed at the beginning of each half stroke and remains attached to the wing until the beginning of the next half stroke. It was found on some small insects, an unsteady inviscid high lift mechanism (clap-and-fling—see Figure 4), consisting on the use of interaction between wings, as they press each other together (like a "clap") at the extreme ends of the stroke, providing a total vertical flapping angle of ~180°.

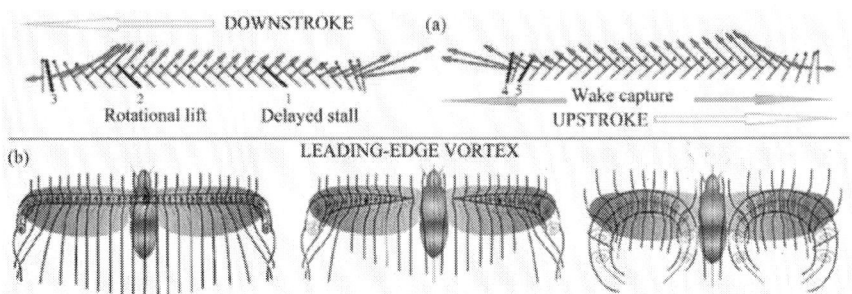

Figure 3. (a) Illustration of a complete insect wingstroke showing the delayed stall and rotational lift on the downstroke movement and the wake capture during all course of the upstroke; blue colour- wing position, with leading edge always on top; red colour-representing the total force [31] ; (b) Illustration of 3 different leading-edge vortex: extends across thorax; attached at the base of the wing and horse-shoe-shaped vortex on both wings.

At the end of the "clap", leading edges began to separate as the trailing edge remains connected initially (V-shape at ~120°); after that, the trailing edge separates as well ("fling"); such movement leads that air to rush into low pressure widening gap and produce high strength vortices of equal and opposite sign [29]. The current state of the art of micro-CT scanners (X-ray Microtomography) is nowadays limited to large insect's wings [30] ; since the resolution of micro-CT scanners is increasing every year, in near future, any insect wing could be successfully scanned. The main challenge for accurately digitizing the wings 3D architecture is minimizing the deformation due to drying of the wing needed to reduce scatter and noise in the scan due to evaporation. Researchers used this technology to scan a dragonfly "Sympetrum vulgatum" forewing and hindwings; they found that both vein and membrane thickness increases from tip to root on both wings, which allows the wing to effectively bear both inertial and aerodynamic loads. On their model, they discover that the inertial loads along the wingspan were approximately 1.5 to 3 times higher than the aerodynamic loads—wings deformation were dominated by inertial loads. Based on computations, they also found that wing deformation was smaller during the downstroke, due to structural asymmetry. By the analyses on the work on several researchers, In fact generically, both inertial and aerodynamic force can be the primary cause of wing deformation: in contrast with above statement, and referring also to dragonfly's natural wings, a study concluded that their deformation was mainly due to the aerodynamic forces. Insects have no active control over the wing configuration during flight. The architecture of the wing and the material properties of its element determine how the wing changes their shape in response momentarily to external forces changing, since the wing movement is very complex. It is nowadays a great challenge to researchers, how to build a model with similar properties; of course, another challenge is how to incorporate the wing flexibility into the theoretical model predicting the aerodynamic force during insect flight—both, remains an ongoing challenge to the researchers [32] . As previously referred, leading-edge vortex is the main flight mechanism that allows insects to be able to fly; studies on the unsteady aerodynamics on the flapping wing of a Manduca sexta hawkmoth robotic model (while hovering) were made by the use of computational fluid dynamic (CFD) modelling [33] ; CFD computations revealed a large leading-edge vortex (LEV1) presence during most of the downstroke movement (from the base at ~60/75% of the wing length); as the wing moves towards horizontal position, this structure becomes larger spiralling vortex with strong axial flow at the core, towards the wing tip. Immediately after the middle of the dowsntroke, the LEV breaks down at 75% of the wing length, creating this, a second LEV (LEV2—also revealing a spiral axual flow towards the wing base) between the wing tip and the broken-down position of the first LEV. At the initial supination rotation, LEV2 is pushed off the leading edge due to wing deceleration and the breakdown point of LEV1 moves into half of wing length. During the upstroke a very-small leading-edge vortex (quite 2D structure) appeared at the wing tip and by the time the wing reaches the middle position, this structure extends from wing tip to the wing base. After the upstroke's middle course, the LEV grows rapidly (comparable in size with the dowstroke LEV1) and hence enlarges the negative pressure region. At the later part of downstroke, the LEV breaks don at ~60/70% if the wing length without shedding the tip vortex. During the pronation, the upstroke LEV remained attached to the leading edge of the lower wing surface and a trailing-edge vortex (TEV) was also detected (larger than LEV, lies below it and run from the base to the tip), probably due to wing rotation. Lentink & Dickinson [34] suggested that LEV could be an efficient high-lift mechanism for small and big hovering insects; this suggestion is reenforcing the idea that insect while hovering, more easily might capture the energy left behind in air, by the previous stroke; even more, the insects with four separated wings (part of them as two pairs coupling wings) might benefit their rear pair on the work of forewing's wake flow. A challenge on build insect's wing models approaching the natural wings morphology was made by Tanaka & Wood [35] . They described the fabrication of an artificial insect (hoverfly) wing with a rich set of topological features by micro molding a thermosetting resin; the venation system diameter varied between 50 - 125 µm heights and the corrugation of the wing measured 100 µm. Both solid veins and membrane were simultaneously formed and integrated by a single molding process, by the use of a layered laser ablation technique; each 3D mold were created with 5 µm resolution in height. The replicated wing matched at-scale high precision surface profiles of the natural one, thus enabling parametric experiments of the functional morphology of insect wings. Authors referred that stiffness measured along the natural and model wings on identical values of magnitude. However, nature had adapted insects with wings were stiffness varies ad varies the location region on the wings. Later, on another investigation [36] on the subject of flexural and torsional wing flexibility, they were able to create a rigid wing model (hoverfly) that could produce more lift than the natural one's, thus, in prejudice of maneuverability, a requirement that insects had at their disposal (at their will), at almost 350 million years: the experience to fly in the skies of the Earth. After filming a beetle tethered flight [37] with a high-speed camera, researchers build a model of wing system of identical size of natural wings, on a way to follow the natural performance of: flapping frequency, stroke plane, wing tip path, wing rotation angles and flapping angles; however, their experience demonstrated to them that flapping frequency and wing rotation need an improvement to satisfy the natural mimic, since the positive vertical force achieved was only ~1/5 of the total weight of the system. From all bibliographic revision on birds, bats and insects, all researchers invoked that those flying animals may benefit aerodynamically from the flexibility of their wings—a general idea stating that temporal wing deformation is the basis of force generation. From the design aspect, flexibility may benefit MAVs as well, from several points of view: aerodynamically and lightness of structures. Since insect's flight maneuverability far higher superior to every maneuverability achieved by any man-made flying vehicle, thus, Barata et al. [38] , made an extensive comparative bibliographyc research on the wings motion patterns on several types of insects on resemblant flights, regarding the expoliting of their in-flight basic principles for the acquired performance. They found that the same insect uses their wings in very differently manners, depending, thus, on the maneuver they intend to carry. Every single movement of their wings generates lift (downstroke or upstroke) and the time rate downstroke/upstroke could easily vary at their will. Some of them possess the ability to use the wings in independent ways (differences on wing stroke, wings with different torsions at the same instant, or even one wing used as aerodynamic brake). Despite the flight mechanisms used by insects are not yet fully understood by humans, their replication for use in MAVs will be even more far from being achieved.

Figure 4. (a) Illustration of forewings and hind wings of a dragonfly and their venation structure as referred on [27] and also a pigmented cell known as pterostigma which mass is frequently greater than that of an equivalent area of adjacent wing and its inertia influences the movement of the whole wing movements. Without the pterostigma, the self-exciting vibrations would set in on the wings after a certain critical speed—pterostigma acts as an inertial regulator of wing pitch; (b) Illustration of the "clap-and-fling" flight mechanism: at a certain moment, the wings press together each other (clap); then wing separate as shown (V-shape), generating a low pressure zone between them.

6. CONCLUSION

The current investigation is focused on the mechanisms involved with natural locomotion (thrust and/or lift). Biological systems with interesting applications to Micro Air-Vehicles (MAVs) are generally inspired on flying insects or birds; however, similarly to the aerodynamics of flight, powered swimming requires animals to overcome drag by producing thrust. Commonalities between natural flying and swimming are analyzed together with flow control issues. As it was shown by several researchers on this bibliography work review, the perception of flight performances held by insects and swimming performances held by fishes are not completely understood. All control surfaces present on living aquatic and flying animals (feet, fins, flippers, or wings) are not designed by nature as of rigid materials; instead, they are elastic materials. Insect's wings are morphological wonder (elastic material: every wing motion is a sum of horizontal, vertical and torsional movement), however, what really enables the wings to make enough force for the animal to stay in the air, is the way insects flap them: at a very high angle of attack, creating a structure at the leading edge of the wing, (tornado-like structure) called leading-edge vortex. Researchers investigated dimensionless numbers to study swimming and flight, and their findings were disappointing since it became clear that different points of view exist in the biomechanics field on how to best define and use. So, successful biology-inspired or biomimetic concepts will depend on the understanding of the natural mechanisms especially when they do not agree with the present engineering design principles. An additional difficulty (and a very important one) is the fact that state of art on elastic materials with identical or similar elastic properties of natural insect's wing, does not exist yet.

ACKNOWLEDGEMENTS

The authors express their gratitude to FCT for the funding of the research project: PTDC/EME-MFE/122849/ 2010: Nature—New-biomimetic Aerodynamic Technologies for Undersized Reynolds.

CITE THIS PAPER

Jorge M. M.Barata,Fernando M. S. P.Neves,Pedro A. R.Manquinho,Telmo A. J.Silva, (2016) Propulsion for Biological Inspired Micro-Air Vehicles (MAVs). *Open Journal of Applied Sciences*,**06**,7-15. doi: 10.4236/ojapps.2016.61002

REFERENCES

1. Borelli, G.A. (1680) De Motu Animalium. A. Bernabo, Rome.

2. Pettigrew, J.B. (1874) Animal Locomotion or Walking, Swimming and Flying with a Dissertation on Aeronautics. D. Appleton & Company, New York.

3. Flight, Flying Machines. http://www.1902encyclopedia.com/F/FLI/flight-flying-machines.html

4. Mueller, T.J. and DeLaurier, J.D. (2003) Aerodynamics of Small Vehicles. Annual Review of Fluid Mechanics, 35, 89-111.http:// dx.doi.org/ 10.1146/annurev.fluid.35.101101.161102

5. McMichael, J.M. and Francis, M.S. (1997) Micro Air Vehicles—Toward a New Dimension in Flight. http://fas.org/irp/program/ collect/docs/ mav_auvsi.htm

6. Hylton, T., Martin, C., Tun, R. and Castelli, V. (2012) The DARPA Nano Air Vehicle Program. Proceedings of 50th AIAA Aerospace Sciences Meeting Including the New Horizons Forum and Aerospace Exposition, Nashville, Tennessee, 9-12 January 2012, 8549-8557. http://dx.doi.org/10.2514/6.2012-583

7. von Ellenrieder, K.D., Parker, K. and Soria, J. (2008) Fluid mechanics of flapping wings. Experimental Thermal and Fluid Science, 32, 1578-1589.http://dx.doi.org/10.1016/j.expthermflusci.2008.05.003

8. Evers, J.H. (2007) Biological Inspiration for Agile Autonomous Air Vehicles. In Platform Innovations and System Integration for Unmanned Air, Land and Sea Vehicles (AVT-SCI Joint Symposium). Meeting Proceedings RTO-MP- AVT-146, Paper 15, RTO, Neuilly-sur-Seine, France, 15-1-15-14.

9. Ho, S., Nassef, H., Pornsinsirirak, N., Tai, Y. and Ho, C. (2003) Unsteady Aerodynamics and Flow Control for Flapping Wing Flyers. Progress in Aerospace Sciences, 39, 635-681.http://dx.doi.org/10.1016/ j.paerosci. 2003. 04.001

10. DeLaurier, J.D. and Harris, J.M. (1982) Experimental Study of Oscillating-Wing Propulsion. Journal of Aircraft, 19, 368-373. http://dx.doi.org/ 10.2514/3.44760

11. Tuncer, I.H. and Platzer, M.F. (2000) Computational Study of Flapping Airfoil Aerodynamics. Journal of Aircraft, 37, 514-520. http://dx.doi.org/ 10.2514/2.2628

12. Rozhdestvensky, K.V. and Ryzhov, V.A. (2003) Aerohydrodynamics of Flapping-Wing Propulsors. Progress in Aerospace Sciences, 39, 585-633.http://dx.doi.org/10.1016/S0376-0421(03)00077-0

13. Lentink, D. and Gerritsma, M. (2003) Influence of Airfoil Shape on Performance in Insect Flight. Proceedings of the 33rd AIAA Fluid Dynamics Conference and Exhibit, Orlando, 23-26 June 2003.

14. An, S., Maeng, J. and Han, C. (2009) Thickness Effect on the Thrust Generation of Heaving Airfoils. Journal of Aircraft, 46, 216-222. http://dx.doi.org/10.2514/1.37903

15. Vandenberghe, N., Childress, S. and Zhang, J. (2006) On Unidirectional Flight of a Free Flapping Wing. Physics of Fluids, 18, 014102-1-14102-8.http://dx.doi.org/10.1063/1.2148989

16. Wang, Z.J. (2000) Vortex Shedding and Frequency Selection in Flapping Flight. Journal of Fluid Mechanics, 410, 323-341.http://dx.doi.org/10.1017/S0022112099008071

17. Cebeci. T., Platzer. M., Chen. H., Chang, K.C. and Shao, J.P. (2005) Analysis of Low-Speed Unsteady Airfoil Flows. Edition 2005, Horizons Publishing Inc., Long Beach.

18. Le, T.Q., Ko, J.H., Byun, D., Park, S.H. and Park, H.C. (2010) Effect of Chord Flexure on Aerodynamic Performance of a Flapping Wing. Journal of Bionic Engineering, 7, 87-94.http:// dx.doi.org/10.1016/S1672-6529(09)60196-7

19. Heathcote, S., Wang, Z. and Gursul, I. (2008) Effect of Spanwise Flexibility on Flapping Wing Propulsion. Journal of Fluid and Structures, 24, 183-199.http://dx.doi.org/10.1016/j.jfluidstructs.2007.08.003

20. Lentink, D., Van Heijst, G.F., Muijres, F.T. and Van Leeuwen, J.L. (2010) Vortex Interactions with Flapping Wings and Fins Can Be Unpredictable. Biology Letters, 6, 394-397. http://dx.doi.org/10.1098/rsbl.2009.0806

21. Lentink, D. and Dickinson, M.H. (2009) Biofluiddynamic Scaling of Flapping, Spinning and Translating Fins and Wings. Journal of Experimental Biology, 212, 2691-2704.Http://Dx.Doi.Org/ 10.1242/Jeb.022251

22. Fish, F.E. and Hui, C.A. (1991) Dolphin Swimming—A Review. Mammal Review, 21, 181-195. http://dx.doi.org/10.1111/j.1365-2907.1991.tb00292.x

23. Fish, F.E., Howle, L.E. and Murray, M.M. (2008) Hydrodynamic Flow Control in Marine Mammals. Integrative and Comparative Biology, 48, 788-800.http://dx.doi.org/10.1093/icb/icn029

24. Wolfgang, M.J., Anderson, J.M., Grosenbaugh, M.A., Yue, D.K. and Triantafyllou, M.S. (1999) Near-Body Flow Dynamics in Swimming Fish. Journal of Experimental Biology, 202, 2303-2327.

25. Drucker, E.G. and Lauder, G.V. (2000) A Hydrodynamic Analysis of Fish Swimming Speed: Wake Structure and Locomotor Force in Slow and Fast Labriform Swimmers. Journal of Experimental Biology, 203, 2379-2393.

26. Choi, H., Park, H., Sagong, W. and Lee, S. (2012) Biomimetic Flow Control Based on Morphological Features of Living Creatures. Physics of Fluids, 24, Article ID: 121302.http://dx.doi.org/10.1063/1.4772063

27. Dickinson, M.H., Lehman, F.-O. and Sane, S.P. (1999) Wing Rotation and the Aerodynamic Basis of Insect Flight. Science, 284, 1954-1960.http://dx.doi.org/10.1126/science.284.5422.1954

28. Sun, J. and Bushan, B. (2012) The Structure and Mechanical Properties of Dragonfly Wings and Their Role on Flyability. Comptes Rendus Mécanique, 340, 3-17.http://dx.doi.org/10.1016/j.crme.2011.11.003

29. Ho, S., Nassef, H., Pornsinsirirak, N., Tai, Y.C. and Ho, C.M. (2003) Unsteady Aerodynamics and Flow Control for Flapping Wing Flyers. Progress in Aerospace Sciences, 39, 635-681. http://dx.doi.org/ 10.1016/ j.paerosci.2003.04.001

30. Jongerius, S.R. and Lentink, D. (2010) Structural Analysis of a Dragonfly Wing. Experimental Mechanics, 50, 1323- 1334. http://dx.doi.org/ 10.1007/s11340-010-9411-x

31. Scientific American, The Sciences, Catching the Wake.http://www. scientificamerican.com/article/catching-the-wake/

32. Yin, B. and Luo, H. (2010) Effect of Wing Inertia on Hovering Performance Flapping Wings. Physics of Fluids, 22, Article ID: 111902. http://dx.doi.org/ 10.1063/1.3499739

33. Liu, H., Ellington, C.P., Kawachi, K., van den Berg, C. and Willmott, A.P. (1998) A Computational Fluid Dynamic Study of Hawkmoth Hovering. Journal of Experimental Biology, 21, 461-477.

34. Lentink, D. and Dickinson, M.H. (2009) Rotational Accelerations Stabilize Leading Edge Vortices on Revolving Fly Wings. Journal of Experimental Biology, 212, 2705-2719.http://dx.doi.org/10.1242/jeb.022269

35. Tanaka, H. and Wood, R.J. (2010) Fabrication of Corrugated Artificial Insect Wings Using Laser Micromachined Molds. Journal of Micromechanics and Microengineering, 20, Article ID: 075008. http:// dx.doi.org/10.1088/0960-1317/20/7/075008

36. Tanaka, H., Whitney, J.P. and Wood, R.J. (2011) Effect of Flexural and Torsional Wing Flexibility on Lift Generation in Hoverfly Flight. Integrative and Comparative Biology, 51, 142-150. http://dx.doi.org/10.1093/icb/icr051

37. Nguyen, Q.V., Park, H.C., Goo, N.S. and Byun. D. (2010) Characteristics of a Beetle's Free Flight and a Flapping- Wing System that Mimics Beetle Flight. Journal of Bionic Engineering, 7, 77-86. http://dx.doi.org/10.1016/ S1672-6529(09)60195-5

38. Barata, J.M.M., Neves, F.M.S.P. and Manquinho, P.A.R. (2015) Comparative Study of Wing's Motion Patterns on Various Types of Insects on Resemblant Flight Stages. Proceedings of the AIAA Atmospheric Flight Mechanics Conference, SciTech 2015, Kissimmee, 5-9 January 2015, 828-848.

CHAPTER 5

Effect of Spacecraft Aerodynamics and Heat Shield Characteristics on Optimal Aeroassisted Transfer

Antonio Mazzaracchio, Mario Marchetti

Astronautical, Electrical, and Energetic Engineering Department, Sapienza University of Rome, Rome, Italy

ABSTRACT

A spacecraft designed to operate in a planetary atmosphere must have an adequate heat shield to withstand the high heat fluxes and heat loads that are generated by aerodynamic heating. Very often, the mass of the thermal protection system is a significant fraction of the total mass of the vehicle. In contrast, performing maneuvers in the atmosphere, that would be very costly in terms of propellant consumption if they were performed completely outside of the atmosphere in a classic way, is a very attractive prospective technique. The advantages and disadvantages in terms of total mass spared must be determined. The mission investigated involves an aeroassisted coplanar transfer from a high to a low Earth orbit. The approach uses a combination of three propulsive impulses in space together with an aerodynamic maneuver in the atmosphere. The heat shield adopted is fully ablative, given the expected high values of the entering heat flux. The convenience of the aeroassisted maneuver and the influence of the parameters involved are evaluated in comparison to a conventional Hohmann transfer. In particular, a parametric analysis is performed by varying the following characteristics of the vehicle: aerodynamic efficiency, mass-to-surface ratio, deorbit impulse, and initial altitude of the orbit. The influence of the thermal protection system is examined by assessing the impact of the type of ablative material employed, the thermal safety factor, and the allowable temperature for the adhesive layer on the substructure. The analysis is conducted with a highly representative thermal model by coupling the dynamic and thermal analyses and using a genetic optimizer. The optimization methodology and the thermal model are completely original. The results indicate the importance of choosing low-density ablative materials, of adopting a suitable thermal safety factor, and of choosing high-performance adhesives. The optimal trajectories obtained correspond to a zero second propulsive impulse.

KEYWORDS

Aeroassisted Maneuver; Heat Shield; Optimization; Orbital Transfer; Thermal Protection System

1. INTRODUCTION

One of the future objectives in space activities is to use aeroassisted orbital maneuvers, i.e., maneuvers that are carried out with the aid of the atmosphere, to satisfy the increasingly stringent constraints of the cost of space missions. Depending on the goal of the mission, aeroassisted maneuvers are performed in different ways: aerogravity assist, aerobraking, and aerocapture. To these maneuvers, one must add the re-entry maneuvers and some possible variants such as the skip re-entry technique. These maneuvers are very attractive, although they are only theoretical at this point (except for rare uses in aerobraking), because they can substantially reduce the requirements of a space mission in terms of propulsion and flight time in favor of, among other things, the possibility of housing a larger payload.

Underlying this approach is the possibility of exploiting the presence of the atmosphere of the celestial body around which one wants to operate to lower the overall energy required. In practice, one tries to execute the complete operation with the help of aerodynamics because orbital maneuvers are quite expensive in terms of propulsion, especially those outside the orbital plan. Thus, the design of a spacecraft with specific reference to atmospheric portion of flight must present an efficient aerodynamic configuration; however, the priority constraint of the overall cost of space missions must be satisfied.

From this brief introduction, it is already clear that once shown the technical feasibility of the aeroassisted maneuver, one must evaluate its convenience compared with alternative hypotheses, such as a classical purely propulsive maneuver. In fact, optimizing a spacecraft and its mission, or more specifically its trajectory for a given mission, is always a compromise between the interests of performance, security, and economics, which are almost always in mutual conflict.

Usually, minimizing the mass of the heat shield while respecting the limits of safety is a primary requirement. Indeed, any savings in terms of the mass of the thermal protection system (TPS) can be translated into an increase of the "useful" mass, namely, into an opportunity to accommodate a greater payload or extend other spacecraft subsystems. The problem becomes more complex when considering an aeroassisted maneuver and taking into account the mass of propellant still required during some of the various phases of the maneuver itself.

In this regard, a basic scheme is usually adopted in the literature to perform an aeroassisted maneuver—the use of propulsion only in space together with one or more atmospheric segments of pure aerodynamic flight—which is also considered in this work. Maximizing the benefits at the propulsive level for an aeroassisted maneuver normally requires a more intensive use of the atmospheric phase of flight. Thus, the resulting trajectory touches the denser layers of the atmosphere with longer crossing times. Therefore, a larger TPS with its relative higher mass fraction must be adopted.

A legitimate question that arises at this point is whether the resulting increase in the mass of the TPS, together with the mass of propellant required to enter the atmosphere and then to achieve the final orbit, may override the convenience of the aeroassisted maneuver compared with a classical operation, which is a purely propulsive extra-atmospheric maneuver. It seems evident that this is an optimization problem.

All of the above considerations concerning both the evaluation of the convenience of the aeroassisted approach and the search for the optimal solution and the methodological sphere itself are the basis of the research questions that motivated this work.

An original procedure was carried out jointly with the implementation of software developed by the authors to optimize the aeroassisted orbital maneuvers using a genetic algorithm (GA) with simultaneous evaluation of the optimal configuration of the associated heat shields and the coupling of the dynamic and thermal analyses. The tool was verified by comparing its results with those found in the literature. This tool, because of its level of implementation details, is suitable for the conceptual development stages of a spacecraft and its mission.

The initial intention was to evaluate the influence of various parameters—orbital, aerodynamic, and dimensional—on the feasibility and convenience of the mission. For an introductory analysis of the problem and to evaluate the importance of various factors, coplanar transfer (aerobraking) from a high Earth orbit (HEO) to a low Earth orbit (LEO), both circular, is proposed as a case study. The vehicle is a delta wing spacecraft that is protected by an ablative TPS with uniform thickness. The problem is studied in the absence of constraints on the maximum allowable entering heat flux.

Beyond the presentation of the problem in Section 1, Section 2 describes the model and the optimization procedure. The case study is presented in Section 3, and the relevant results and analyses are discussed in Section 4. Finally, Section 5 offers a summary, conclusions, and recommendations for future improvements.

2. MODELS AND OPTIMIZATION

The description of the models, the governing equations, and the relevant assumptions for the problem—i.e., thermal models, aerodynamic heating, atmospheric flight mechanics, and heat shield configuration—are thoroughly presented in [1] and [2]. Reference [2] can be consulted for full details of the original optimization procedure.

All of the analysis presented here was performed using the cited software developed by the authors called ATHSHO (Aeroassisted Trajectory and Heat SHield Optimization). It is important to recall that thermal analysis is performed with a one-dimensional plane model and that the adopted GA refers to a mixed one-point/two-point crossover operator together with a reproduction plan that provides a full generational replacement with elitism [3-5].

3. CASE STUDY

A parametric analysis was performed by varying the following characteristics of the vehicle: aerodynamic efficiency, mass-to-surface ratio, deorbit impulse, and the altitude of the initial orbit. The influence of the characteristics of the TPS was examined by assessing the type of ablative material adopted, the thermal safety factor, and the allowable temperature for the bond-line, i.e., the adhesive layer on the substructure.

3.1. Hypotheses

The main hypotheses considered are the following:

1. The initial total mass of the vehicle is given.

2. The attitude control is accomplished only through the angle of attack.

3.   The entering heat flux is unconstrained.

4.   The TPS is fully ablative with uniform thickness.

3.2. Vehicle

The vehicle model is a delta wing shuttle with a high L/D ratio, which is comparable in the first instance to the configuration and dimensions of the Boeing X-37A vehicle (**Figure 1**).

The dimensions, sizes, and aerodynamic characteristics of the vehicle used were taken in part from [6-8]. Other data were reasonable assumptions made by the authors. The main dimensions and characteristics of the vehicle are listed in **Table 1**, and the principal aerodynamic and propulsive parameters are listed in **Table 2**.

The values shown are those considered for the nominal reference case. Some of them may vary depending on the purpose of the study (Section 4).

Figure 1. Boeing X-37A (source NASA.gov).

Table 1. Vehicle dimensions and characteristics.

Vehicle length	l_{ve}	9.38 m
Vehicle body radius	r_b	1.00 m
Vehicle wing span	ws_{ve}	4.50 m
Vehicle wing cord	wc_{ve}	3.50 m
Vehicle reference surface	S	11.69 m^2
Vehicle TPS total surface	$S_{TPS,ve}$	42.65 m^2
Bond-line limit temperature	$T_{BL,lim}$	450 K
Thermal safety factor	TSF	1

Table 2. Vehicle's aerodynamic and propulsive characteristics.

Zero-lift drag coefficient	C_{D0}	0.032
Induced drag factor	K_D	1.4
Lift coefficient derivative	$C_{L,\alpha}$	0.5699
Maximum lift coefficient	$C_{L,max}$	0.4
Propellant specific impulse	I_{sp}	310 s

3.3. Mission and Maneuver

Table 3 lists the values of the altitudes for the initial Geostationary Earth Orbit (GEO) and final LEO for the aerobraking maneuver as well as the conventional height assumed for the atmosphere.

Even in this case, the indicated values are those used for the reference case. In particular, in the parametric analysis performed, the value of the initial HEO altitude was varied. The physical properties of the atmosphere were derived from the model 1976 US Standard Atmosphere [9].

Figure 2 shows a classic schematic for a HEO-LEO aerobraking maneuver. The strategy involves the combined use of aerodynamic maneuvering in the atmosphere and some extra-atmospheric propulsion phases. More precisely, one assumes that the propulsive phases are concentrated in three impulses in space and that the portion of atmospheric flight is performed without the use of the propulsion system. The first propulsive impulse (deorbit) at the altitude H_A of the initial HEO varies the vehicle's speed by ΔV_1 to enter the atmosphere along an elliptic orbit segment. The second impulse (boost) is applied upon the exit from the atmosphere to achieve the final LEO by ascending once more along an elliptic orbit segment. The boost is expressed by a speed variation of the vehicle, ΔV_2.

Table 3. Maneuver characteristics.

Initial HEO (GEO) altitude	H_A	35,786 km
Final LEO altitude	H_B	480 km
Atmosphere's upper limit	H_{am}	129.6 km

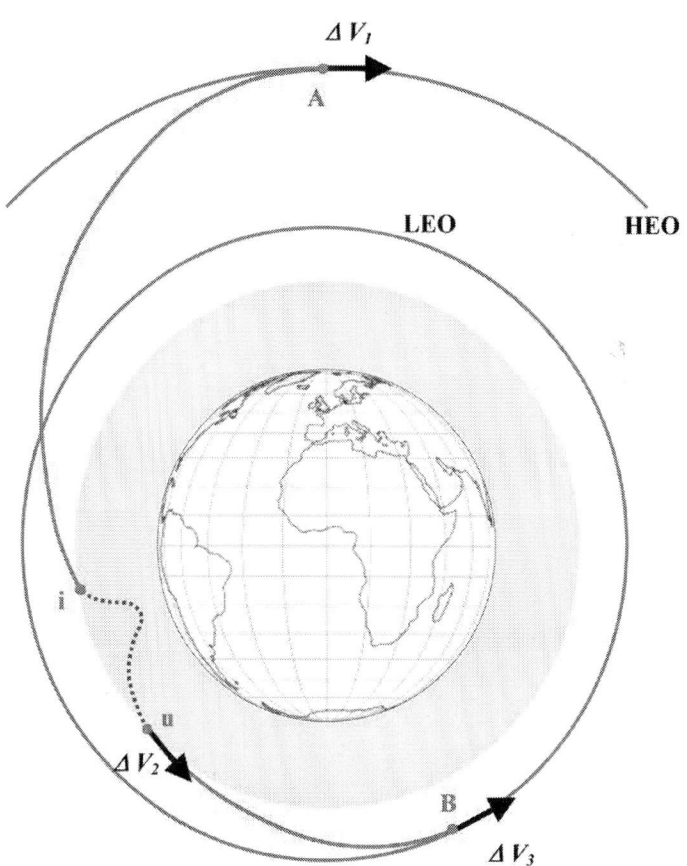

Figure 2. Schematic of a HEO-LEO aerobraking maneuver.

The third and final impulse (circularizing) varies the speed by ΔV_3 to circularize the vehicle's path within the altitude H_B of the final LEO.

It is useful at this point to describe the various phases of the maneuver in more detail as follows. Initially, the vehicle is moving on a circular orbit of radius R_A with a speed V_A around the Earth, which has radius $R_\oplus$. The expression of the circular speed is the following:

$$V_A = \sqrt{\mu/R_A} \tag{1}$$

Where

$$R_A = R_\oplus + H_A. \tag{2}$$

The deorbit is accomplished by applying the first impulse ΔV_1 in the opposite direction of the spacecraft's speed. This impulse puts the vehicle along an elliptic orbit with the perigee inside the dense layers of the atmosphere.

The atmospheric region below the altitude H_{atm} in which the aerodynamic effects are considered to be conventionally present is denoted the sensible atmosphere. One can determine the speed V_i and the flight path angle γ_i of the vehicle's trajectory at the atmospheric entry point ($H = H_{atm}$) through the following relationships, which are obtained according to energetic considerations:

$$V_i = \sqrt{2\mu \left(\frac{1}{R_{atm}} + \frac{(V_A - \Delta V_1)^2}{2} - \frac{1}{R_A} \right)} \tag{3}$$

$$\gamma_i = \cos^{-1} \left[\frac{R_A}{R_{atm} V_i} (V_A - \Delta V_1) \right] \tag{4}$$

Where

$$R_{atm} = R_{\oplus} + H_{atm}.$$

(5)

Clearly, it is necessary that the applied be greater than the minimum $\Delta V_{1,min}$ for which there would be only a tangential trajectory to the edge of the sensible atmosphere. This $\Delta V_{1,min}$ is presented in **Figure 3** and given by the following expression:

$$\Delta V_{1,min} = \sqrt{\frac{\mu}{R_A}} - \sqrt{2\mu \frac{\dfrac{1}{R_{atm}} - \dfrac{1}{R_A}}{\left(\dfrac{R_A}{R_{atm}}\right)^2 - 1}}.$$

(6)

During the atmospheric portion of flight, the vehicle performs the required maneuver, which is optimally controlled by modulations of the angle of attack α (leaving the bank angle σ fixed) subject to the heating constraints, if any. During this phase, the vehicle's speed decreases because of aerodynamic drag, and because of this loss of energy, a new impulse is necessary to achieve the final altitude.

At the end of atmospheric flight, the vehicle is situated at an altitude H_{atm} again, is driven at a speed V_u, and has a flight path angle equal to γ_u. At this moment, a boost impulse is applied to enter an ascending elliptic orbit with the apogee equal to the radius of the final circular orbit.

The required ΔV_2, as a function of V_u and γ_u, can be found from the following expression:

$$\Delta V_2 = \sqrt{2\mu \frac{\dfrac{1}{R_B} - \dfrac{1}{R_{atm}}}{1 - \left(\dfrac{R_B}{R_{atm}}\right)^2 \cos^2 \gamma_u}} - V_u$$

(7)

Where

$$R_B = R_\oplus + H_B.$$

(8)

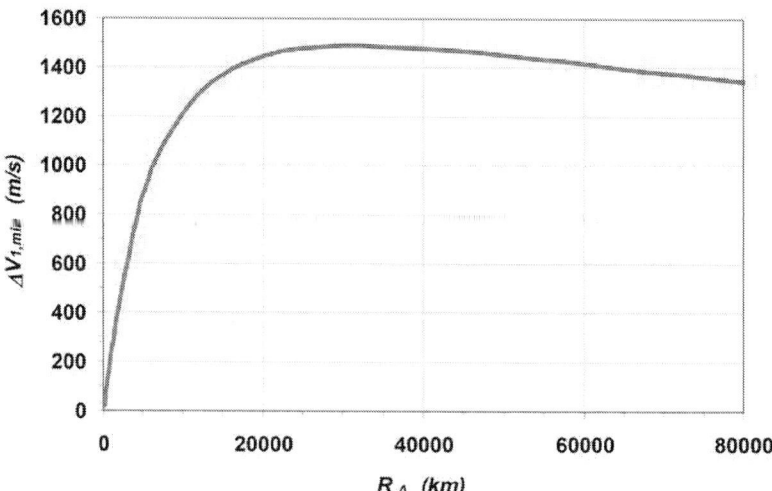

Figure 3. $\Delta V_{1,min}$ vs initial HEO radius.

Once the final altitude is reached, the third impulse is applied to circularize the final orbit. The expression for the third ΔV is the following:

$$\Delta V_3 = \sqrt{\frac{\mu}{R_B}} - \sqrt{2\mu \frac{\dfrac{1}{R_B} - \dfrac{1}{R_{atm}}}{1 - \left(\dfrac{R_B}{R_{atm}}\right)^2 \cos^2 \gamma_u}} \cdot \frac{R_B}{R_{atm}} \cos \gamma_u.$$

(9)

3.4. Fitness Function, Objective Function, and Constraints

The total initial mass of the vehicle is the sum of the propellant mass, the TPS mass, and the structural and payload masses. The goal of the current optimization problem is to perform the assigned orbital transfer while minimizing the sum of the mass of the propellant and the mass of the TPS needed. The objective function, which must be maximized, is then given by the final mass of the vehicle, which can be defined as the performance index of the problem. There are some events that cause variation in the mass of the vehicle. The first is the consumption of fuel due to the deorbit impulse; thus, one can calculate the mass of the vehicle entering the atmosphere. The latter can be derived directly using the Tsiolkovsky equation for the impulse in question:

$$m_{ve,i} = m_{ve,ini} \, e^{-\frac{|\Delta V_1|}{g_0 \, I_{sp}}} . \tag{10}$$

This mass is further reduced by $m_{TPS,los}$ during the passage through the atmosphere because of TPS ablation. The mass loss due to ablation comes from both the surface recession and the material density change due to pyrolysis. Thus, the mass of the vehicle at the atmosphere exit is the following:

$$m_{ve,u} = m_{ve,i} - m_{TPS,los} . \tag{11}$$

At this point, the boost and circularization impulses are applied in sequence, and the final vehicle mass is obtained by two successive applications of the Tsiolkovsky equation:

$$m_{ve,fin} = m_{ve,u} \, e^{-\frac{|\Delta V_2|+|\Delta V_3|}{g_0 \, I_{sp}}} . \tag{12}$$

Compliance with the eventual constraint on the entering heat flux is ensured through a reward factor that is added to the objective function ($m_{ve,fin}/m_{ve,ini}$) by means of an appropriate multiplicative weight w_{HF}, which is chosen specifically to achieve rapid convergence. Likewise, the objective function is multiplied by its weight w_m. Thus, the final expression for the fitness function for the genetic optimizer is as follows:

$$ff = w_m \frac{m_{ve,\,fin}}{m_{ve,\,ini}} + w_{HF} R_{f,\,HF}.$$

(13)

In this case study, the heat flux is considered unconstrained, and assuming $w_m = 1$, the previous expression reduces to the following:

$$ff = \frac{m_{ve,\,fin}}{m_{ve,\,ini}}.$$

(14)

That is, the fitness function coincides with the objective function.

3.5. Assessment of the Convenience

The convenience of the aeroassisted maneuver can be assessed with respect to an equivalent Hohmann transfer. Such a transfer is completely "all propulsive" outside the atmosphere and based on two impulsive velocity changes at two points with radii R_A and R_B, respectively:

$$\Delta V_A = \sqrt{\frac{\mu}{R_A}} \left(\sqrt{\frac{2 R_B}{R_A + R_B}} - 1 \right)$$

(15)

$$\Delta V_B = \sqrt{\frac{\mu}{R_B}} \left(1 - \sqrt{\frac{2 R_A}{R_A + R_B}} \right).$$

(16)

Thus, the total "all propulsive" impulse is:

$$\Delta V_{ap} = \left|\Delta V_A\right| + \left|\Delta V_B\right|.$$

(17)

Consequently, the expression for the final mass of the vehicle in the case of the Hohmann transfer is as follows:

$$m_{ve,fin,ap} = m_{ve,ini}\, e^{-\frac{\left|\Delta V_{ap}\right|}{g_0\, I_{sp}}}.$$

(18)

4. RESULTS AND ANALYSIS

The case study for the HEO-LEO coplanar transfer through aerobraking, which was described in the previous Section 3, has been addressed by parameterising the mass-tosurface ratio M/S of the vehicle, the ΔV_1 corresponding to the deorbit impulse applied, the aerodynamic efficiency ε (with $C_{L,max}$ fixed), the ablative material used, the maximum allowable temperature for the bondline $T_{BL,lim}$, and the thermal safety factor TSF. In addition, the above parametric analysis was performed for different values of the initial HEO altitude, which are expressed as fractions of the GEO.

More specifically, the following values were adopted:

- HEO Altitude: 35,786 km (=GEO); 53,679 km (HEOH = 1.5 GEO); 17,893 km (HEOL = 0.5 GEO). The altitude for the final LEO is the same for all cases and equal to 480 km.

- M/S: 200 kg/m², 300 kg/m² and 450 kg/m², corresponding, respectively, to low, medium and high load values. Incidentally, 300 kg/m² is the value found most frequently in the literature. These load values correspond to the initial mass of the vehicle: 2338.0 kg, 3507.0 kg, and 5260.5 kg.

- ΔV_1 range: the range of the parametric first variation of V (in steps of 10 m/s) between the minimum value (rounded to the nearest ten higher) given by Equation (6) and the maximum value that allows a solution to complete the aeroassisted maneuver. The minimum values of the deorbit impulse for the three HEO altitudes analyzed are (**Figure 3**) 1490 m/s for GEO, 1440 m/s for HEOH, and 1420 m/s for HEOL.

- ε : 1.5 and 3 correspond, respectively, to a medium and a high aerodynamic efficiency. These values were obtained by varying C_{D0} and K_D.

- TFS : 1 and 2 correspond, respectively, to a minimum nominal thickness of the TPS and a "safer" double thickness.

- $T_{BL,\,lim}$: 176.85°C (450 K) and 400°C (673.15 K) were both acquired from data in the literature; the first is used for standard adhesives, and the second is used for high-performance adhesives.

- Existing adhesives with enhanced characteristics allow the threshold temperature to be increased to 400°C, which results in greater heat shield thickness savings [10].

- Ablative material: PICA-15 (hereinafter referred to as PICA), AVCOAT 5026-HCG (hereinafter referred to as AVCOAT), and FM 5055 CP in its Reduced Density version (hereinafter referred to as RDCP).

Among all the possible combinations of these parameters with their associated values, 12 cases were analyzed for each value of the ratio for a total of 36 different scenarios. The combinations analyzed are presented in Tables 4-6, where the index (a, b, and c) marks the adopted value for the load. Moreover, because the range of variation of is analyzed by steps of 10 m/s, 390 different cases form the database of results for the analysis of the case study in question.

As a reference, the nominal case is chosen for the GEOHEO transfer that provides a TPS made of PICA, with = 450 K, TSF = 1, and ε = 1. The twelve combinations chosen and the corresponding purpose of the evaluation are summarized in **Table 7**. Figures 4 to 15 report the trends of the mass gain compared with the "all propulsive" case with varied applied and parameterized as a function of the three chosen values of M/S, which were analyzed for each scenario.

The graph on the left of each figure shows the cited mass gain in absolute terms, whereas the graph on the right presents its performance in terms of the percentage of the initial vehicle's mass.

Table 4. HEO-LEO transfer, parameters and values (cases 1, 2, 3, and 4).

		Case 1a	Case 1b	Case 1c	Case 2a	Case 2b	Case 2c	Case 3a	Case 3b	Case 3c	Case 4a	Case 4b	Case 4c
HEO-LEO													
HEO = GEO	35,786 km - 480 km	✳	✳	✳	✳	✳	✳	✳	✳	✳	✳	✳	✳
M/S													
200 kg/m²	$M_{s,bi}=2338.0$ kg	✳			✳			✳			✳		
300 kg/m²	$M_{s,bi}=3507.0$ kg		✳			✳			✳			✳	
450 kg/m²	$M_{s,bi}=5260.5$ kg			✳			✳			✳			✳
Aerodynamic efficiency													
Medium 1.5	$C_{D0}=0.1; K=1.111$	✳	✳	✳	✳	✳	✳				✳	✳	✳
High 3.0	$C_{D0}=0.017; K=1.76$							✳	✳	✳			
ΔV₁													
	1490 - 1590 m/s	✳											
	1490 - 1560 m/s		✳										
	1490 - 1520 m/s			✳									
	1490 - 1590 m/s				✳								
	1490 - 1550 m/s					✳							
	1490 - 1520 m/s						✳						
	1490 - 1710 m/s							✳					
	1490 - 1600 m/s								✳				
	1490 - 1560 m/s									✳			
	1500 - 1600 m/s										✳		
	1490 - 1560 m/s											✳	
	1490 - 1520 m/s												✳
Ablative material													
	PICA-15	✳	✳	✳	✳	✳	✳	✳	✳	✳	✳	✳	✳
T-bond line													
	450.00 K	✳	✳	✳				✳	✳	✳	✳	✳	✳
	673.15 K				✳	✳	✳						
Thermal safety factor													
	1	✳	✳	✳	✳	✳	✳	✳	✳	✳			
	2										✳	✳	✳

Table 5. HEO-LEO transfer, parameters and values (cases 5, 6, 7, and 8).

		Case 5a	Case 5b	Case 5c	Case 6a	Case 6b	Case 6c	Case 7a	Case 7b	Case 7c	Case 8a	Case 8b	Case 8c
HEO-LEO													
HEO = GEO	35,786 km - 480 km	✳	✳	✳	✳	✳	✳						
HEOH = 1.5 GEO	53,679 km - 480 km							✳	✳	✳	✳	✳	✳
M/S													
200 kg/m²	$M_{s,bi}=2338.0$ kg	✳			✳			✳			✳		
300 kg/m²	$M_{s,bi}=3507.0$ kg		✳			✳			✳			✳	
450 kg/m²	$M_{s,bi}=5260.5$ kg			✳			✳			✳			✳
Aerodynamic efficiency													
Medium 1.5	$C_{D0}=0.1; K=1.111$	✳	✳	✳	✳	✳	✳	✳	✳	✳	✳	✳	✳
ΔV₁													
	No solution	✳											
	No solution		✳										
	1500 - 1640 m/s			✳									
	1490 - 1570 m/s				✳								
	1490 - 1520 m/s					✳							
	1490 - 1500 m/s						✳						
	1440 - 1490 m/s							✳					
	1440 - 1470 m/s								✳				
	1440 - 1460 m/s									✳			
	1440 - 1480 m/s										✳		
	1440 - 1460 m/s											✳	
	1440 - 1450 m/s												✳
Ablative material													
	PICA-15							✳	✳	✳			
	FM 5055 RDCP	✳	✳	✳									
	AVCOAT 5026-H				✳	✳	✳				✳	✳	✳
	CG												
T-bond line													
	450.00 K	✳	✳	✳	✳	✳	✳	✳	✳	✳	✳	✳	✳
Thermal safety factor													
	1	✳	✳	✳	✳	✳	✳	✳	✳	✳	✳	✳	✳

The mass gain is therefore a measure of the convenience of the aeroassisted maneuver compared with the corresponding Hohmann transfer. When reading these graphics, the definitions of the "range of feasibility" and the "range of convenience" are introduced.

Table 6. HEO-LEO transfer, parameters and values (cases 9, 10, 11, and 12).

		Case 9a	Case 9b	Case 9c	Case 10a	Case 10b	Case 10c	Case 11a	Case 11b	Case 11c	Case 12a	Case 12b	Case 12c
HEO-LEO													
HEOH = 1.5 GEO	53,679 km - 480 km	※	※	※	※	※	※						
HEOL = 0.5 GEO	17,893 km - 480 km							※	※	※	※	※	※
M/S													
200 kg/m²	$M_{ve,tot}$ = 2338.0 kg	※			※			※			※		
300 kg/m²	$M_{ve,tot}$ = 3507.0 kg		※			※			※			※	
450 kg/m²	$M_{ve,tot}$ = 5260.5 kg			※			※			※			※
Aerodynamic efficiency													
Medium 1.5	C_{D0} = 0.1; K = 1.111							※	※	※			
High 3.0	C_{D0} = 0.017; K = 1.76	※	※	※	※	※	※				※	※	※
ΔV_1													
	1440 - 1530 m/s	※											
	1440 - 1490 m/s		※										
	1440 - 1460 m/s			※									
	1440 - 1520 m/s				※								
	1440 - 1470 m/s					※							
	1440 - 1450 m/s						※						
	1420 - 1880 m/s							※					
	1420 - 1640 m/s								※				
	1420 - 1550 m/s									※			
	1420 - 1880 m/s										※		
	1420 - 1810 m/s											※	
	1420 - 1620 m/s												※
Ablative material		※	※	※				※	※	※	※	※	※
	AVCOAT 5026-HCG				※	※	※						
T-bond line													
	450.00 K	※	※	※	※	※	※	※	※	※	※	※	※
Thermal safety factor													
	1	※	※	※	※	※	※	※	※	※	※	※	※

The first one is the interval of in which the aeroassisted maneuver is feasible, whereas the second interval is the range in which the aeroassisted operation is more convenient than the "all propulsive" one. As mentioned above, having made the run for all values of leading to a solution, the range of feasibility is directly readable as the x-axis interval of the definition of the curves themselves.

The range of convenience is by definition the interval in which the curve takes positive values. The range of convenience is always within the range of feasibility. **Table 8** shows, for the value M/S = 200 kg/m² (case "a" of Tables 4-6), the values of the three propulsive impulses applied in each scenario, which in each case give the highest convenience for the maneuver (maxima of the "blue" curves in the graphs of Figures 4 to 15).

It is evident that the utmost convenience for the aeroassisted maneuver corresponds to a value of that is very close to zero, i.e., in the absence of the boost impulse (case 5a has no feasible solutions for the considered value for the load).

This outcome is valid in general for the other values of . Thus, the trajectory optimization process tends to reduce the vehicle's speed while exiting the atmosphere to the "right" value to reach the final LEO with minimal energy expenditure. Saving this fuel for the second impulse is the basis for the achievement of the highest convenience.

Comparing Figures 4 to 15, one can draw some interesting conclusions:

- The increase of   in all cases involves a decrease in the range of feasibility; in particular, the maximum possible   diminishes (the right bound of the range).

- The increase of   (**Figure 5**) from the standard value to the high-performance value due to the reduced thickness of the TPS improves the maximum achievable convenience (by approximately 2%).

- The increase in the aerodynamic efficiency of the vehicle (**Figure 6**) involves a significant extension of the range of feasibility, but a more moderate increase in the range of convenience. Cases with higher   benefit most. Even here, the maximum achievable convenience increases (approximately 2% more).

- The doubling of the thermal safety factor (**Figure 7**), which increases the mass of the TPS, represents a significant performance penalty for the aeroassisted maneuver, which remains affordable for a short interval of   only for the case with highest   .

- The use of other ablative materials (Figures 8 and 9), compared with the reference material PICA, appears to be inappropriate for a GEO initial altitude. Using the RDCP, the sole case with the highest   is feasible, but still not convenient. Conversely, using the AVCOAT increases the feasibility compared with RDCP but limits convenience compared with PICA. Thus, the relatively low density of the PICA is a discriminating element in the selection of the ablative material. Actually, these results are conservative because of the assumption that the vehicle is covered with a TPS with a uniform initial thickness.

- In **Figure 8**, all runs with load values of 200 kg/m^2 and 300 kg/m^2 give results, but the curves are not represented because the final "useful" masses (i.e., all the masses that are not propellant and TPS) are negative. Thus, the vehicle is able to get out of the atmosphere, but without enough fuel to complete the mission.

- Figures 10 (HEO high altitude) and 14 (HEO low altitude) show that the aeroassisted maneuver is much more convenient if the HEO altitude is higher. Lowering the initial altitude reduces the convenience but considerably extends the range of feasibility. The reduced convenience at low altitudes is due to the fact that the Hohmann transfer requires a lower fuel quantity with the lowered the initial altitude for the same final LEO. In contrast, for the corresponding aeroassisted maneuver, even if the TPS is less consistent, the propulsive contribution remains significant.

- Figures 10-13, considered in pairs (the first two and the second two), provide a performance comparison between PICA and AVCOAT, respectively, for high initial orbits in the case of low and high aerodynamic efficiency. The comparison is in favor of the use of PICA, albeit less marked than for GEO orbits. For high orbits for a vehicle with high ε, the performance of PICA remains considerable; however, the performance of AVCOAT, although lower than PICA, is notable. In this context, it seems appropriate to comment on NASA's recent evaluation concerning the choice of material for ablative TPS of the Orion spacecraft. The final decision was between PICA and AVCOAT, and NASA ultimately chose the latter. PICA has been used to date only on a re-entry capsule of the Stardust probe, while AVCOAT has a long history of reliability (Apollo missions). Though the results of the present work provide evidence of the tangible prevalence of the PICA performance, the reliability of AVCOAT most likely outweighs these performance considerations.

- The increase in aerodynamic efficiency has less impact in the case of lower orbits (Figures 14 and 15).

Table 7. HEO-LEO transfer, scenarios analyzed.

	Combination	Purpose
1	GEO; PICA; ε medium; $T_{BL,low}$ standard; TSF = 1	Reference nominal case
2	GEO; PICA; ε medium; $T_{BL,low}$ high performance; TSF = 1	$T_{BL,low}$ influence
3	GEO; PICA; ε high; $T_{BL,low}$ standard; TSF = 1	ε influence
4	GEO; PICA; ε medium; $T_{BL,low}$ standard; TSF = 2	TSF influence
5	GEO; RDCP; ε medium; $T_{BL,low}$ standard; TSF = 1	Material influence
6	GEO; AVCOAT; ε medium; $T_{BL,low}$ standard; TSF = 1	Material influence
7	HEOH; PICA; ε medium; $T_{BL,low}$ standard; TSF = 1	HEO altitude influence; reference case for HEOH
8	HEOH; AVCOAT; ε medium; $T_{BL,low}$ standard; TSF = 1	Material influence w.r.t. HEOH
9	HEOH; PICA; ε high; $T_{BL,low}$ standard; TSF = 1	ε influence w.r.t. HEOH
10	HEOH; AVCOAT; ε high; $T_{BL,low}$ standard; TSF = 1	Material + ε influence w.r.t. HEOH
11	HEOL; PICA; ε medium; $T_{BL,low}$ standard; TSF = 1	HEO altitude influence; reference case for HEOL
12	HEOL; PICA; ε high; $T_{BL,low}$ standard; TSF = 1	ε influence w.r.t. HEOL

Legend: in red: GEO nominal case and changes w.r.t. it; in blue: HEOH nominal case and changes w.r.t. it; in green: HEOL nominal case and changes w.r.t. it.

Table 8. Propulsive impulses in the more convenient maneuvers for "a" cases.

Case	ΔV_1(m/s)	ΔV_2 (m/s)	ΔV_3 (m/s)
1a	1530.00	1.38	334.75
2a	1530.00	-1.83	332.91
3a	1500.00	-2.21	375.26
4a	1520.00	0.69	526.53
5a	-	-	-
6a	1520.00	-0.95	502.54
7a	1480.00	2.86	366.44
8a	1470.00	-4.00	568.53
9a	1450.00	-3.51	423.53
10a	1450.00	0.67	421.13
11a	1460.00	-3.56	410.11
12a	1440.00	-2.73	267.69

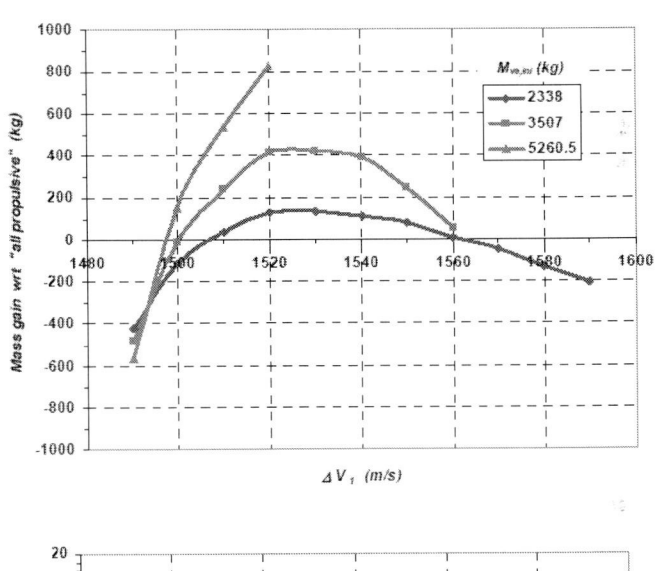

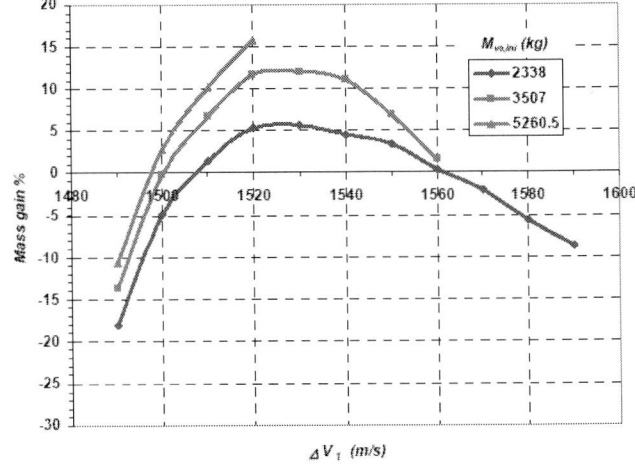

Figure 4. HEO-LEO scenario 1: reference nominal case.

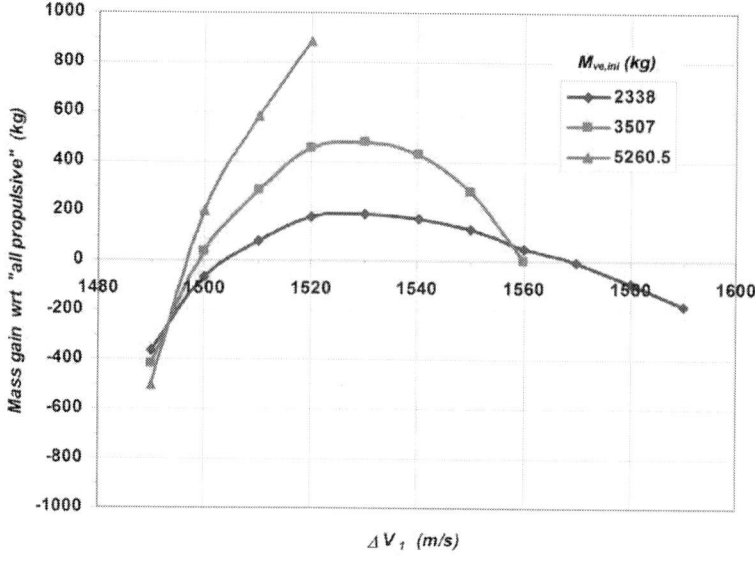

Figure 5. HEO-LEO scenario 2: $T_{BL,lim}$ influence.

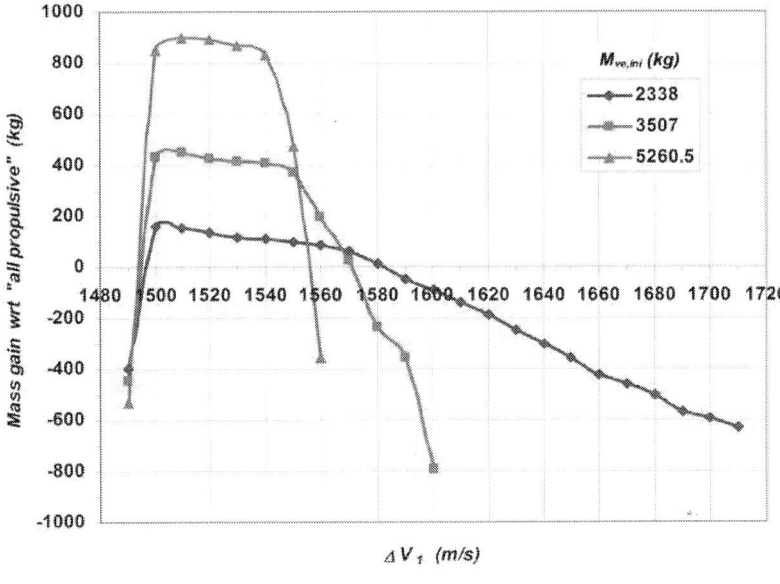

Figure 6. HEO-LEO scenario 3: e influence.

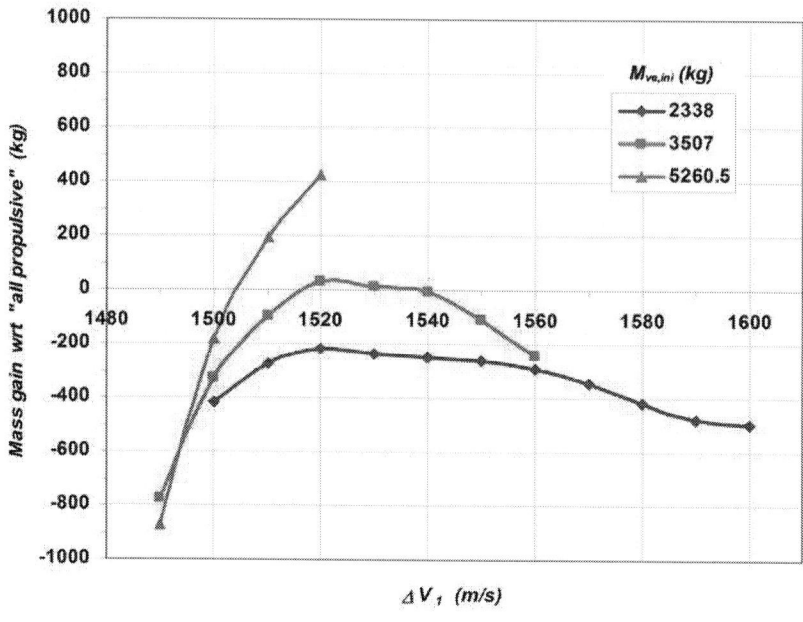

Figure 7. HEO-LEO scenario 4: TSF influence.

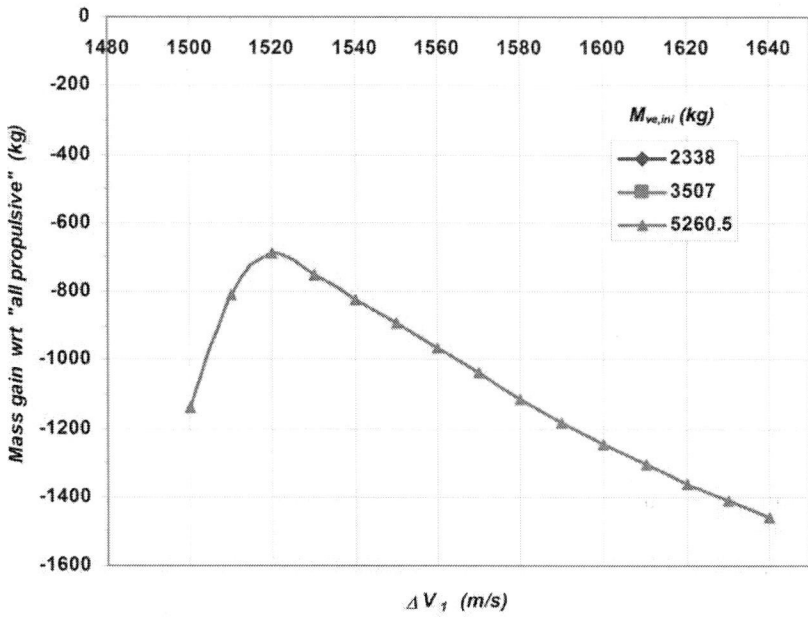

Figure 8. HEO-LEO scenario 5: material (RDCP) influence.

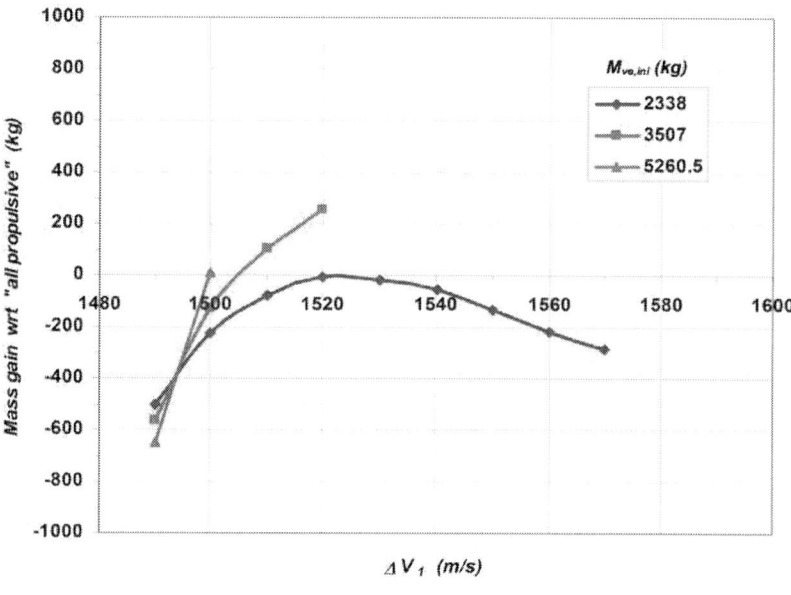

Figure 9. HEO-LEO scenario 6: material (AVCOAT) influence.

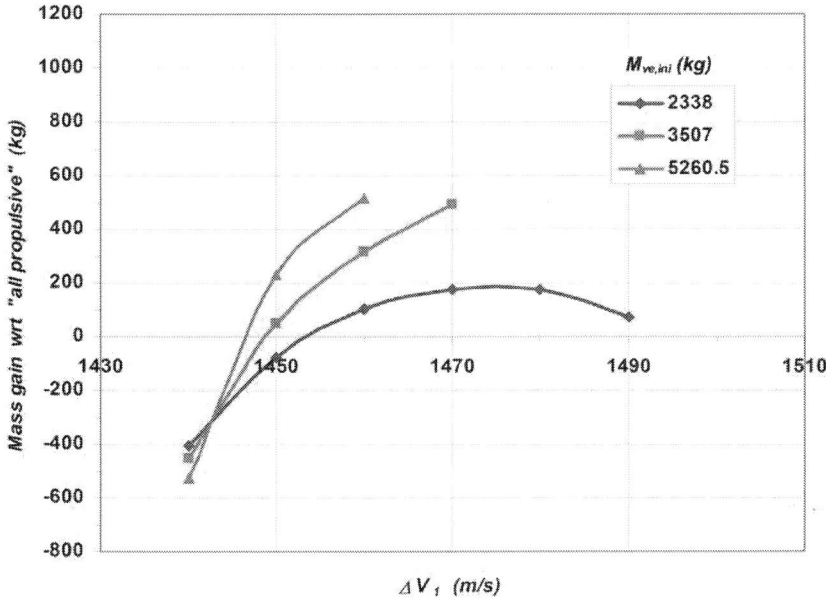

Figure 10. HEO-LEO scenario 7: HEO altitude influence; reference case for HEOH.

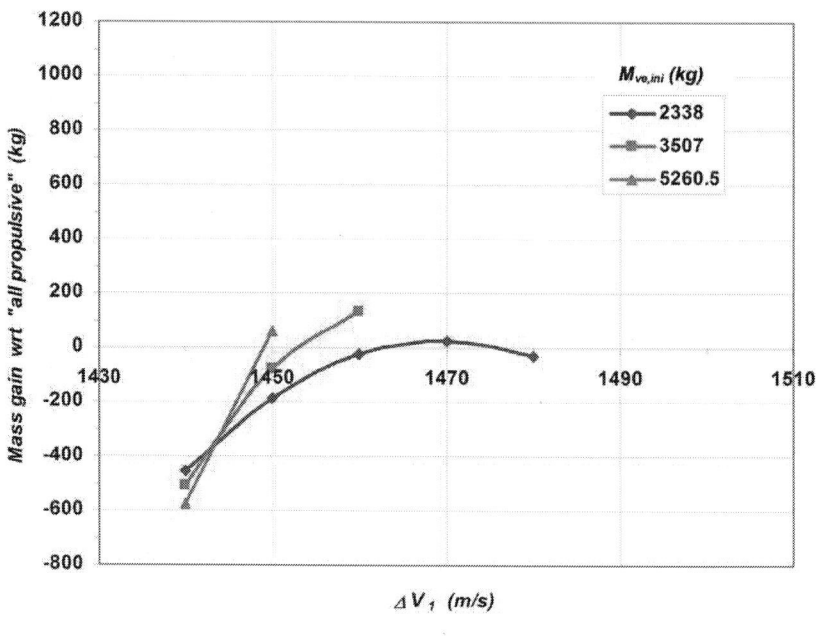

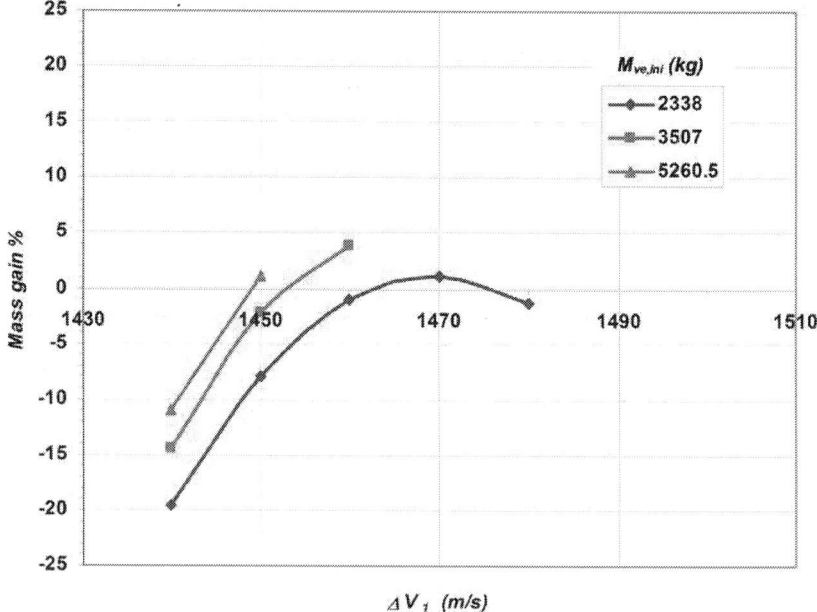

Figure 11. HEO-LEO scenario 8: material (AVCOAT) influence w.r.t. HEOH.

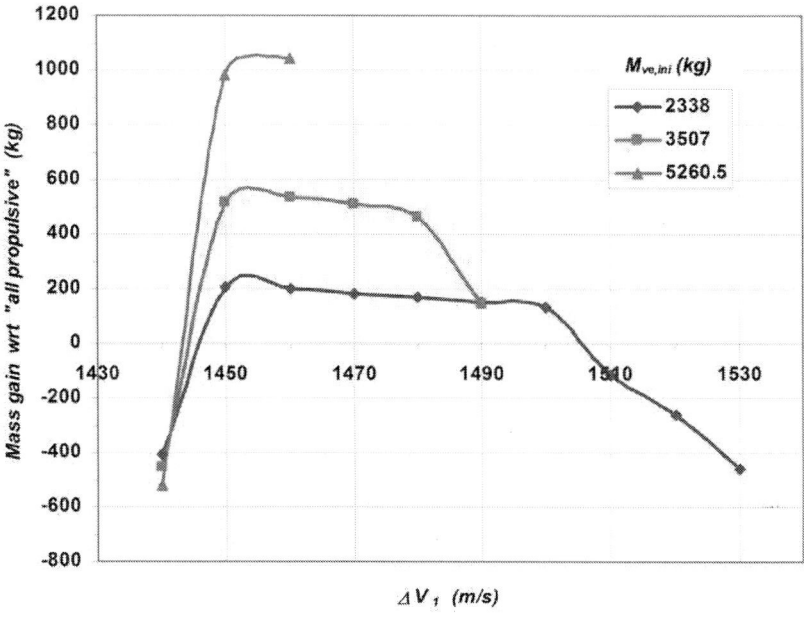

Figure 12. HEO-LEO scenario 9: e influence w.r.t. HEOH.

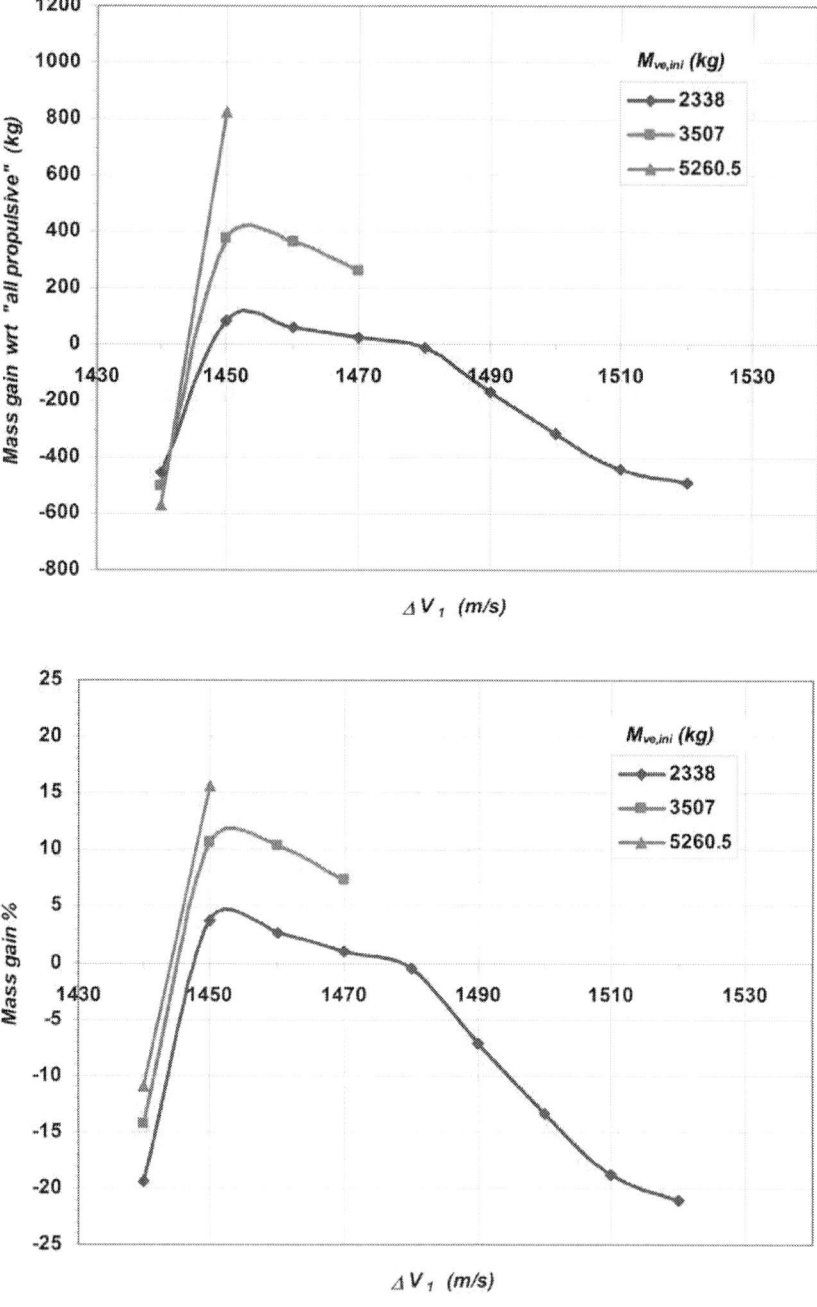

Figure 13. HEO-LEO scenario 10: Material (AVCOAT) + e influence w.r.t. HEOH.

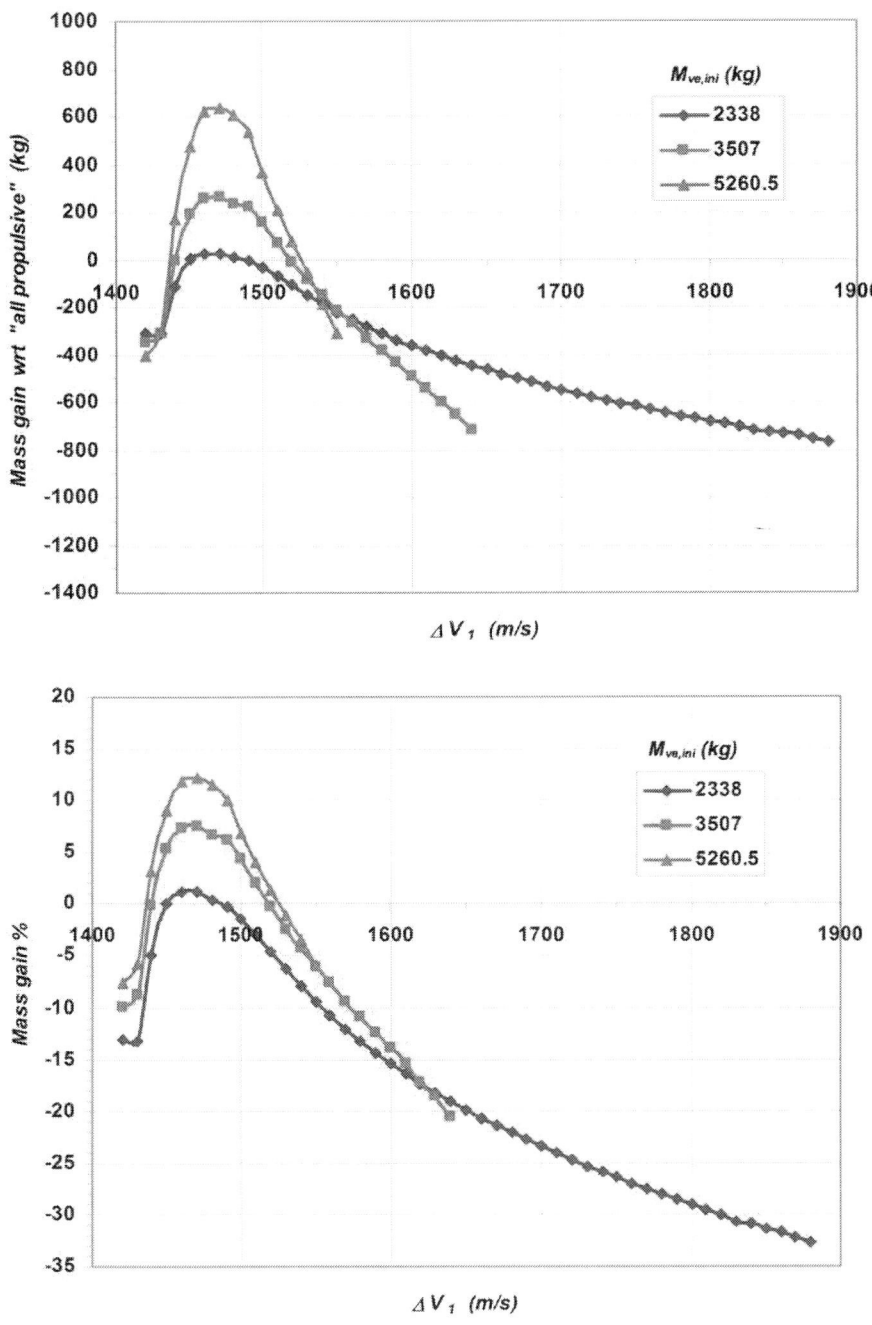

Figure 14. HEO-LEO scenario 11: HEO altitude influence; reference case for HEOL.

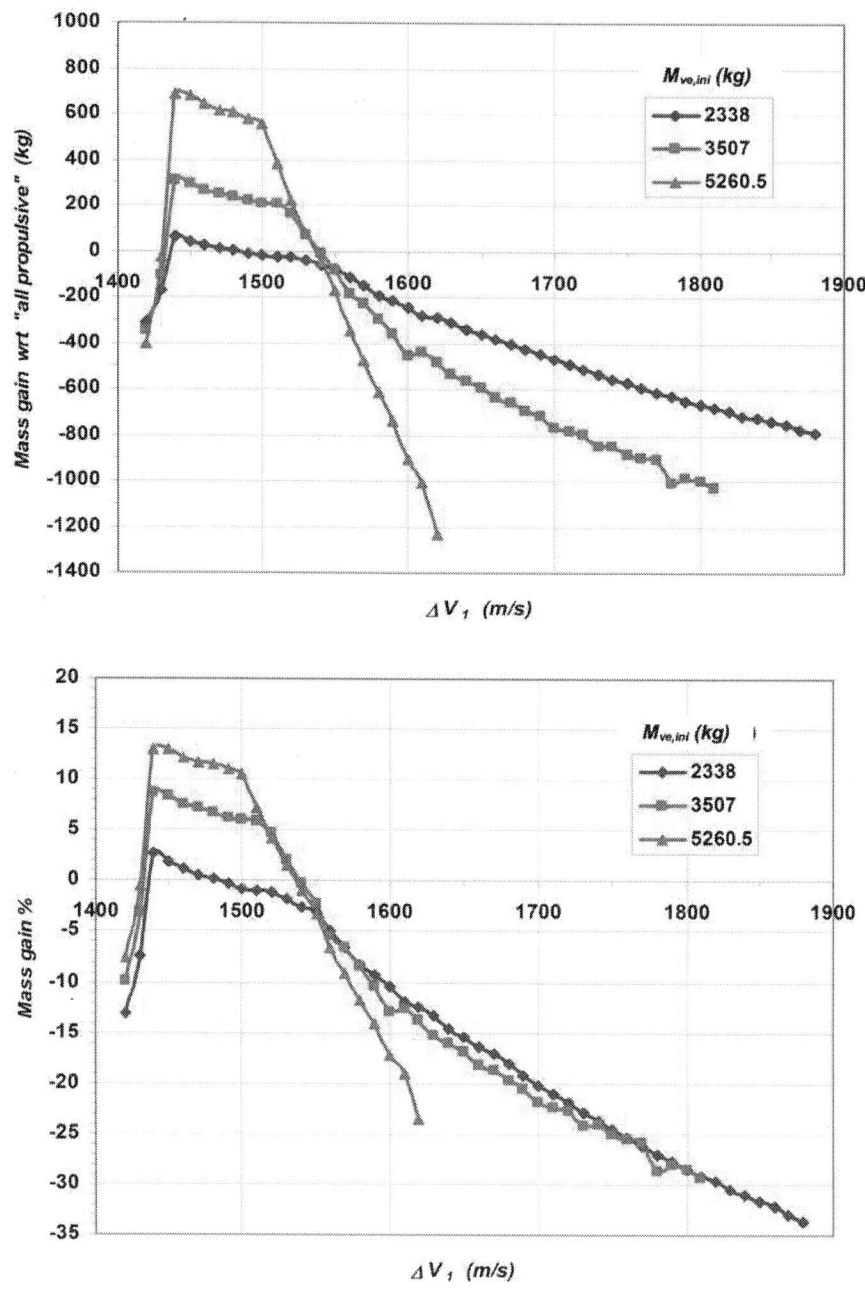

Figure 15. HEO-LEO scenario 12: e influence w.r.t. HEOL.

5. SUMMARY AND CONCLUSIONS

Some aeroassisted transfers between coplanar orbits were analyzed to examine the influence of the aerodynamic characteristics of a spacecraft and the characteristics of its heat shield on the maneuver's final performance.

The study was conducted with an original tool developed by the authors comprising highly representative models for thermal analysis, atmospheric flight dynamics, and an optimizer based on a genetic algorithm. The convenience of the maneuver was evaluated by comparison with the corresponding extra-atmospheric Hohmann transfer with two impulses.

The results of the case study indicated that there is a range of feasibility for the maneuver, depending on the first propulsive impulse of deorbit, within which there is, in turn, an interval in which the maneuver is more convenient than the homologous Hohmann transfer in terms of the total mass savings achieved (TPS and propellant). In particular, the results show that in terms of the convenience of the aeroassisted maneuver, the influence of the ablative material type used (low-density advantage) and of the aerodynamic efficiency is the most relevant, whereas other characteristics play a minor role. Significant benefits can be obtained from a careful thermal design, with a suitable thermal safety factor and high-performance adhesives for the bond-line.

For all cases, the highest convenience, and consequently the corresponding optimal trajectories, is obtained when it is not necessary to apply the second propulsion impulse. The most important future development for this type of analysis is increasing the number of cases analyzed in terms of materials, types of maneuvers and spacecraft characteristics.

REFERENCES

1. A. Mazzaracchio and M. Marchetti, "A Probabilistic Sizing Tool and Monte Carlo Analysis for Entry Vehicle Ablative Thermal Protection Systems," Acta Astronautica, Vol. 66, No. 5-6, 2010, pp. 821-835. doi: 10.1016/j.actaastro.2009.08.033

2. A. Mazzaracchio and M. Marchetti, "Coupled Aeroassisted Orbital Plane Change Manoeuvre and Thermal Protection System Optimisation," 61st International Astronautical Congress, Prague, September 27-October 1, 2010.

3. P. Charbonneau and B. Knap, "A User'S Guide to Pikaia 1.0," Boulder, Colorado, 1995.

4. P. Charbonneau, "An Introduction to Genetic Algorithms for Numerical Optimization," Boulder, Colorado, 2002.

5. P. Charbonneau, "Release Notes for Pikaia 1.2," Boulder, Colorado, 2002.

6. C. Gogu, T. Matsumura, R. T. Haftka and A. V. Rao, "Aeroassisted Orbital Transfer Trajectory Optimization Considering Thermal Protection System Mass," Journal of Guidance, Control and Dynamics, Vol. 32, No. 3, 2009, pp. 927-938.doi:10.2514/1.37684

7. NASA Marshall Space Flight Center, "X-37 Demonstrator to Test Future Launch Technologies in Orbit and Reentry Environments," NASA Facts, May 2003, FS-2003-05- 65-MSFC.

8. Y. Y. Shi and D. H. Young, "Minimum Fuel Coplanar Aeroassisted Orbital Transfer Using Collocation and Nonlinear Programming," Flight Mechanics/Estimation Theory Symposium, NASA Goddard Space Flight Center, 1991, pp. 461-480 (SEEN92-1407005-13).

9. US Standard Atmosphere, US Government Printing Office, Washington DC, 1976.

10. T. J. Collins, W. M. Congdon, S. S. Smeltzer and K. S. Whitley, "High-Temperature Structures, Adhesives, and Advanced Thermal Protection Materials for Next-Generation Aeroshell Design," NASA Langley Research Center, 2006, Paper 2M-02-2005.

Chapter 6

Numerical Study of the Influence of Elements inside the Wheelhouse on the Passenger Vehicle Aerodynamic

Edvaldo Angelo[1], Gabriel Angelo[2,3], Pedro Henrique Di Giovanni Santos[3], Delvonei Alves de Andrade[3]

[1]*Grupo de Simulação Numérica (GSN), Universidade Presbiteriana Mackenzie, São Paulo, Brazil*

[2]*Department of Mechanical Engineering, Centro Universitário da Fundação Educacional Inaciana Padre Sabóia de Medeiros (FEI), São Bernardo do Campo, Brazil*

[3]*Instituto de Pesquisas Energéticas e Nucleares, IPEN CNEN/SP, Universidade de São Paulo, São Paulo, Brazil*

ABSTRACT

Models for the study of computational fluid dynamics in vehicles to determine aerodynamic loads usually take into account only the geometry of the body. Several constructive elements such as the wheel geometry or suspension components are disregarded in the computational models. This work presents the study of the aerodynamics of a one-fourth model passenger vehicle, which contains the wheelhouse interior elements. The goal is to identify the aerodynamic loads produced by these components and their effect on the flow dynamics. Wheel and tire set, brake components, suspension and drive shaft are contemplated. Computer simulations were performed to the vehicle speed varying from 0 to 120 km/h and included the rotation of the tire and wheel assembly, considering the tire geometry in dynamic conditions. The computational model is solved by the finite volume method, wherein the computational domain is divided into tetrahedral and hexahedral elements. The turbulence model used is the standard $k - \varepsilon$.

KEYWORDS

Computational Fluid Dynamics, Wheelhouse, Aerodynamics, Vehicle Aerodynamics

1. INTRODUCTION

The numerical study for the calculation of aerodynamic forces arising in a vehicle has become more conventional. The methods usually employed for the analysis of such forces in vehicles are designed after the devise of prototypes in wind tunnels, which are very expensive demanding a large amount of equipment, facilities and skilled labor.

Currently companies have performed three-dimensional construction of their models, thus facilitating processes of tools construction, assembly interference checking and also allowing the numerical analysis.

Most of the articles related to numerical studies of vehicle aerodynamics aim at the vehicle body [1] or phenomena associated with lateral wind [2] generally disregarding components as the wheel geometries, tire, brake assembly (disc and caliper), suspension, shock absorber, stub axle and wheelhouse.

This article aims to determine the flow influence on a passenger vehicle using a one-fourth model of a complete vehicle as seen in Figure 1.

Many researchers have been trying to understand the complex flow patterns around vehicles, developing measurement techniques in wind tunnels [3] and computational efforts [4] . Few of the studies [5] , however, are concentrated in the region of wheelhouse. Generally, the studies include the complete vehicle and indirectly study the region of the wheelhouse.

The aim of this study is to identify, through numerical simulations, the effects of internal elements of the wheelhouse, the wheelhouse itself, the wheel and tire assembly in drag force of a passenger vehicle, through the modelling of a quarter car.

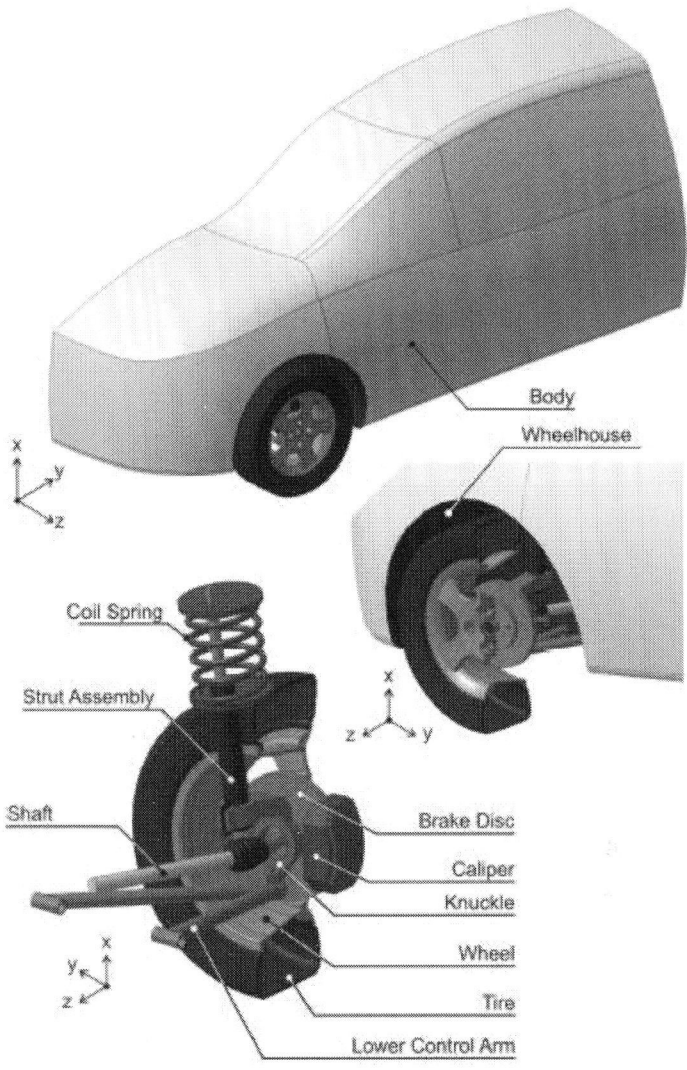

Figure 1. One-fourth model and components to be evaluated.

2. MATHEMATICAL MODEL AND BOUNDARY CONDITIONS

The finite volume method was used to solve the conservation equations applied to an unstructured mesh. The equations solved for the model are respectively, the mass and the momentum conservation equations. It is used the standard k –

ε model [6] for the turbulence treatment and the fluid is air at a temperature (T) of 25°C. The flow regime is transient, varying the inlet velocity (v_{in}) of the fluid and the soil (v_g) from zero to 33.5 m/s ($v_{in} = v_g$). Mach number (M) for the proposed condition in this analysis is less than 0.3 so the flow can also be considered as incompressible [7] and thermal effects are neglected, i.e. isothermal flow. The wheel angular velocity (w) can be expressed as a function of ground speed and the dynamic tire radius (r_d) [8] . The boundary conditions are: non-slip flow for the vehicle surfaces, free-slip flow on the top and side walls and symmetry for the zx plane as presented in Figure 2.

A simplification of some elements was adopted for the proposed model given that other geometries exert little influence on the flow. These elements are the hood of the constant velocity joint, wheel bolts case, braking set details, etc. The respective changes can be observed in Figure 3.

The approach presented by Stern and Wilson [9] [10] to define and verify the mesh is used in this paper. They discuss the mesh dependency on the results focusing the element size definition in order to validate the CFD models. The methodology considers an increase of the mesh density for the same boundary condition using predefined ratios. This procedure must be performed in such a way that property variation or small variations are not present. When this condition is satisfied the solution is considered independent of the mesh. The computational domain for the last refinement iteration process has 2.8 million elements. In Figure 4(a) it is possible to observe the surface elements applied to the vehicle body. A greater density of elements has been applied inside the wheelhouse and the tire. Figure 4(b) and Figure 4(d), and an auxiliary view is presented for the volume density elements verification, Figure 4(c).

3. RESULTS AND DISCUSSION

The numerical study was performed in transient regime. In order to demonstrate the highest gradients of the quantities involved, air input speed was chosen to 33.5 m/s (120 km/h). The pressure variation as a function of the model surface is indicated on the color map of Figure 5(a). Sections in the computational domain were constructed and can be observed in Figure 5(c). Figure 5(b) shows a view (cut) in the xy plane and height (z) equal to r_d. This section region is protected by the vehicle body which contributes to a small pressure variation against the other variations. The sections in the xz plane, Figure 5(d) and Figure 5, detail A, show respectively, the pressure gradient in the section and on the surface geometry tire/wheel. Also in Figure 5 the frontal location of the vehicle shows stagnation regions, this behavior indicates consistency in the numerical results [11] .

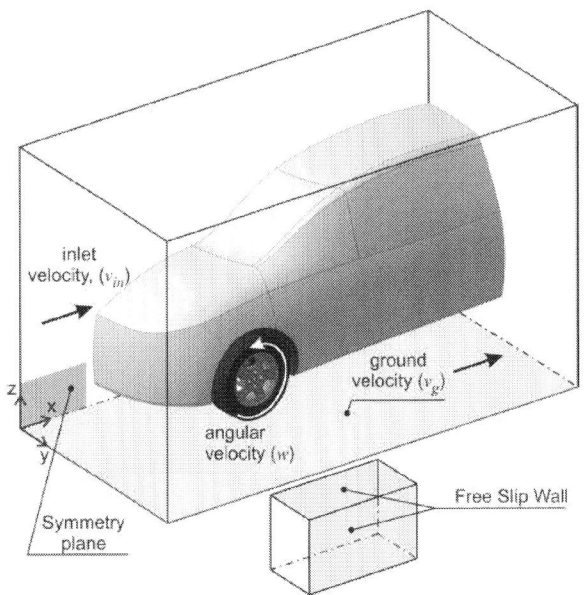

Figure 2. Boundary conditions for the numerical model.

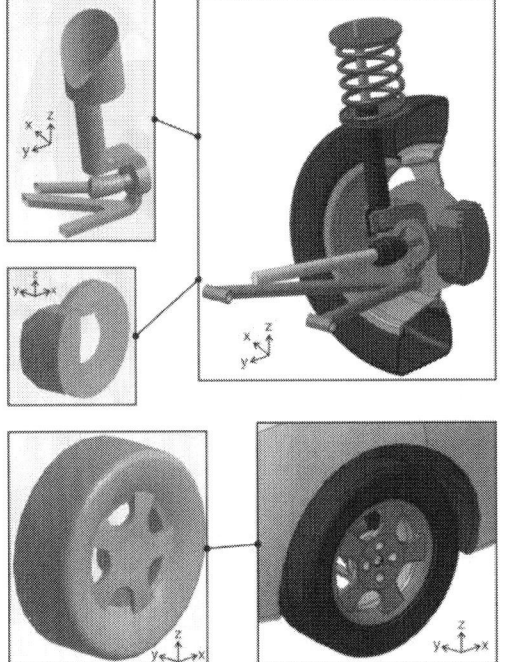

Figure 3. Geometrical simplifications.

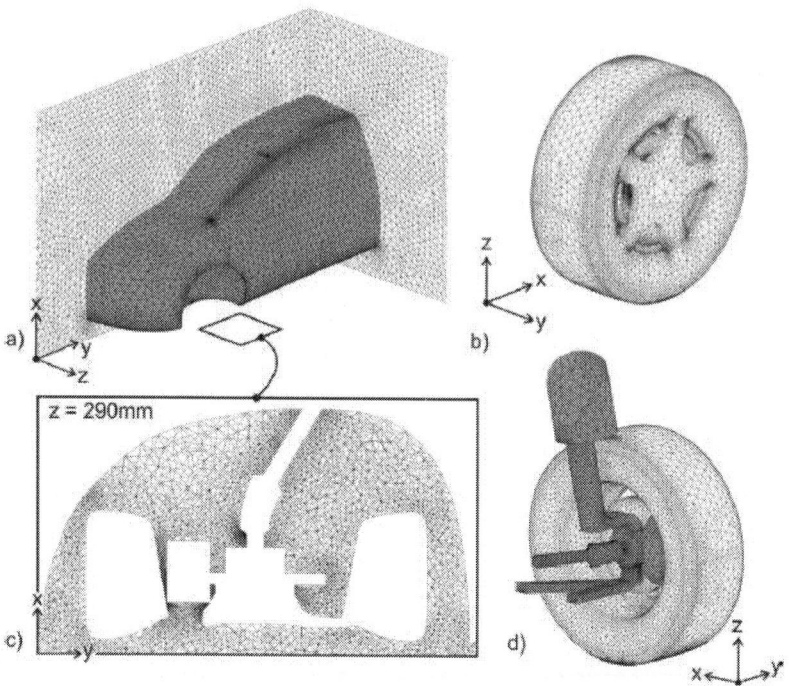

z = 290mm

Figure 4. (a) Triangular elements in the region of the vehicle body; (b) Surface elements applied on the wheel geometry; (c) Detail view in the xy plane of the wheelhouse region; (d) Superficial elements inside the wheelhouse.

In Figure 6 streamlines colored by velocity gradient were constructed in order to make evident the flow behavior. Figure 6 also shows that the vehicle external geometry does not originates vortex structures. This vortex structures occur only in the lower region of the wheel, tire and wheelhouse.

Vortex structures appeared depending on the geometry of the wheelhouse as shown in Figure 7. This characteristic in some cases depending on the vortex frequency emission can excite surface so as to promote noise. However, this simulation does not take into consideration such factors. Another problem associated with the formation of these vortex inside the wheelhouse is the decrease of the air flow that cools the brake system. Thus in the wheelhouse design these recirculation structures should be avoided [12] .

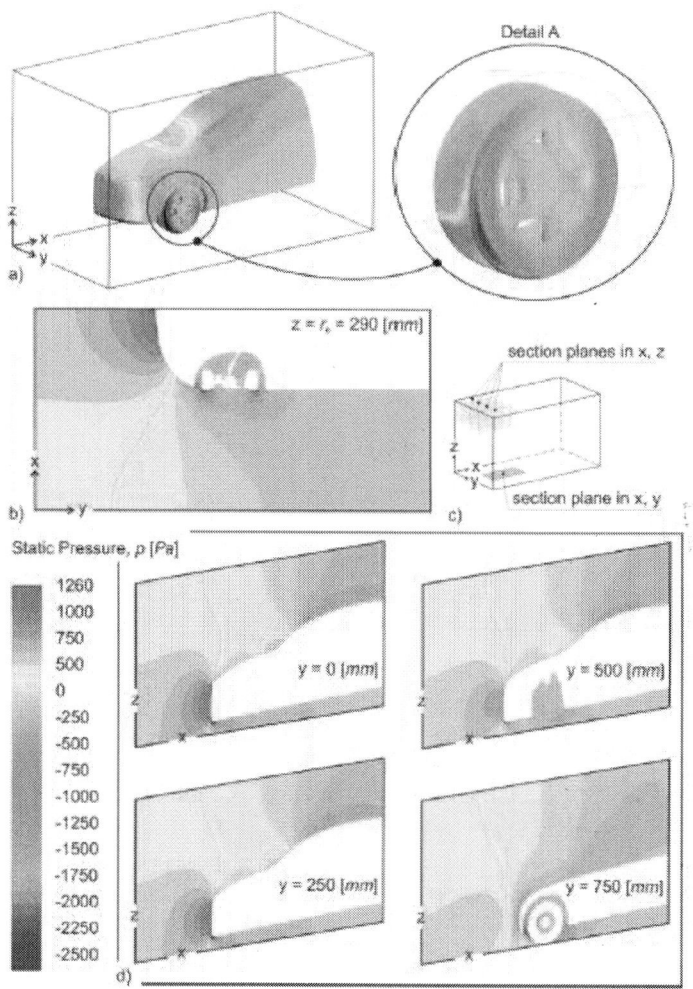

Figure 5. (a) Pressure gradient applied to the vehicle surface; (b) Pressure gradient on xy plane at axis z equal to r_d; (c) Illustration showing the positions of sections; and (d) Pressure gradient applied to xz plane at various positions.

The aerodynamic drag force (F_a) according to Fox and McDonald [7] is defined by Equation (1):

$$F_a = C_d \cdot \rho \cdot A_f \cdot \frac{v^2}{2}$$

$$(1)$$

where: C_d—aerodynamic drag coefficient, ρ—density and A_f—frontal area.

The aerodynamic drag coefficient is not constant [13] . It depends directly on the flow characteristics. Thus the Reynolds number is defined by Equation (2):

$$Re = \frac{\rho \cdot v \cdot L}{\mu}$$

(2)

where: L is the is the characteristic length chosen as $L = \dfrac{A_f}{P_f}$ and P_f perimeter as function of the frontal area.

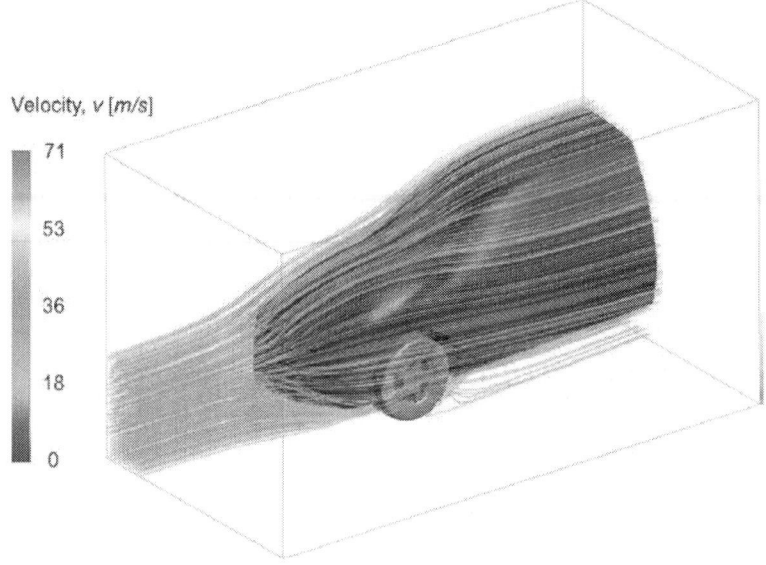

Figure 6. Streamlines in the proximity of the vehicle surface.

The respective areas and perimeters for this model can be observed in Figure 8.

The numerical simulations and experimental results reported in the literature indicate that the drag coefficient values vary slightly from a given speed [14] . In such cases, the value stabilizes, even with the increased speed and allows it to be defined as aerodynamic characteristics of the body. Gillespie, 1992 [8] presents average values tables for the drag coefficient as a function of the vehicle category which is proper only for estimates.

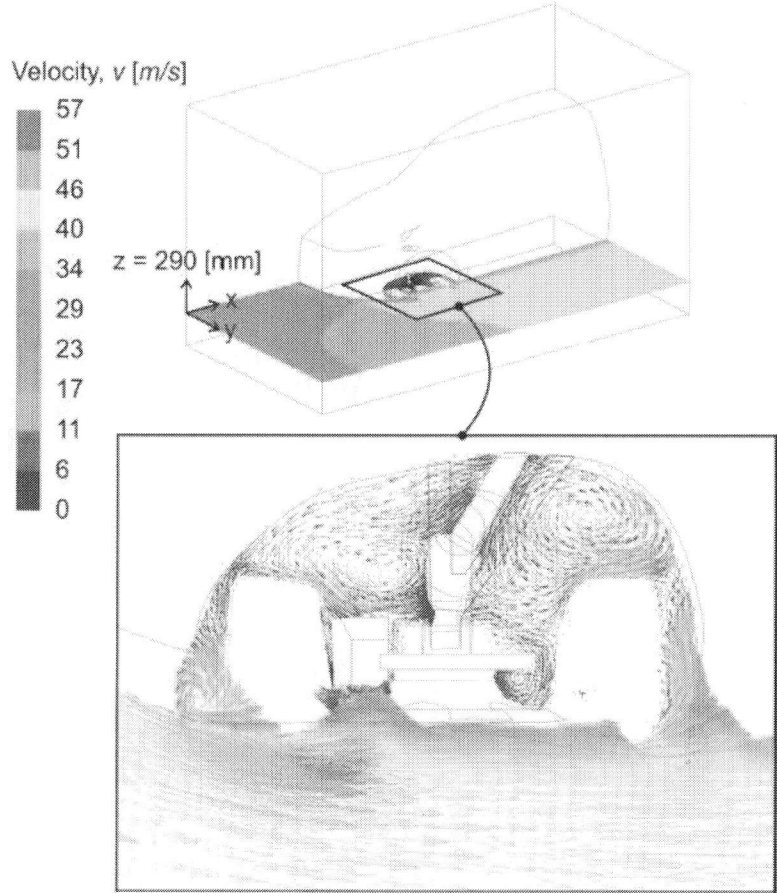

Figure 7. Cross section of the gradient and velocity vectors on the xy plane and z = 290 mm.

For the verification of such feature a graph of velocity versus drag coefficient was built. The vehicle body and the whole vehicle body have different drag coefficients, but show the same trend as shown in Figure 9.

Figure 10 presents the drag force projected in the x direction as a function of velocity. As expected the greatest strength comes from the vehicle body, nevertheless the geometry effect of the tire, wheel and wheelhouse are significant and together come to 16.3% of the total drag force in this direction at a speed of 33.5 (m/s) and is directly proportional to the velocity.

The internal elements of the wheelhouse exert a small influence on the total drag force of the vehicle. It is about 0.4%, so that adopted simplification compared with the order of magnitude of other elements is very much reasonable.

4. CONCLUSIONS

The main conclusions of the study are:

1) It is possible the use of mathematical models for predicting aerodynamic loads;

2) Flow inside the wheelhouse is complex and presents recirculation regions;

3) Mechanical elements within the wheelhouse had simplified geometry and were incorporated into the model; however, the contribution to the drag force of such elements is quantitatively less than that of the wheelhouse and assembly wheel and tire;

4) According to the simulations, presented in paper, neglecting the effects of the box on wheels and wheel tire set is not suitable in aerodynamic vehicle research, except at low speeds where the aerodynamic loads are low for any vehicle.

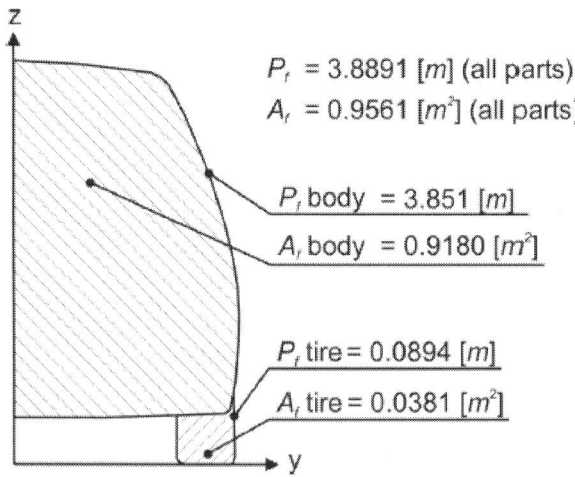

$P_f = 3.8891 \ [m]$ (all parts)
$A_f = 0.9561 \ [m^2]$ (all parts)

$P_f \ \mathrm{body} = 3.851 \ [m]$
$A_f \ \mathrm{body} = 0.9180 \ [m^2]$

$P_f \ \mathrm{tire} = 0.0894 \ [m]$
$A_f \ \mathrm{tire} = 0.0381 \ [m^2]$

Figure 8. Vehicle frontal projection.

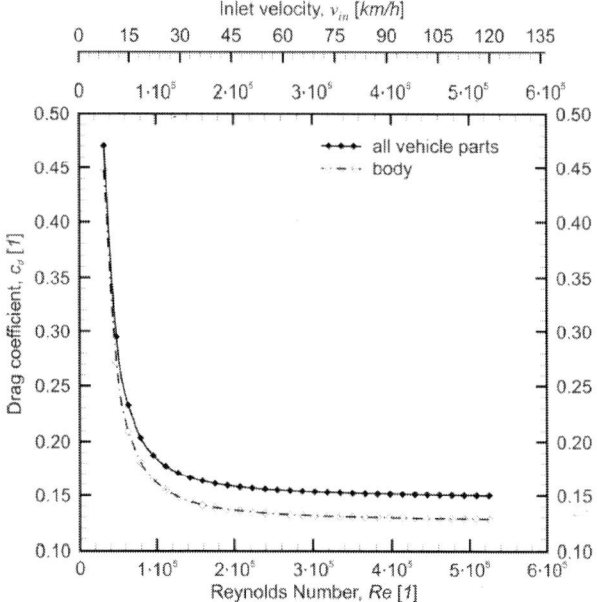

Figure 9. Inlet velocity and Reynolds number as a function of the drag coefficient.

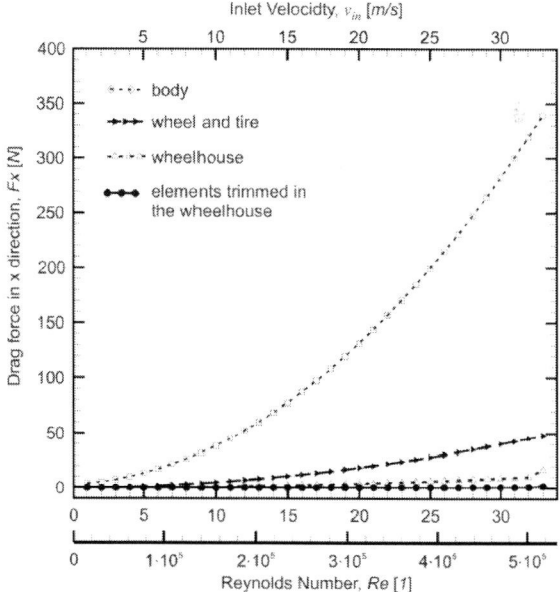

Figure 10. Inlet speed as a function of the strength's projection in the x direction.

REFERENCES

1. Beaudoin, J.-F. and Aider, J.-L. (2008) Drag and Lift Reduction of a 3D Bluff Body Using Flaps. Experiments in Fluids, 44, 491-501.

2. Corin, R.J., He, L. and Dominy, R.G. (2008) A CFD Investigation into the Transient Aerodynamic Forces on Overtaking Road Vehicle Models. Journal of Wind Engineering and Industrial Aerodynamics, 96, 1390-1411.

3. Cogotti, A. (2008) Evolution of Performance of an Automotive Wind Tunnel. Journal of Wind Engineering and Industrial Aerodynamics, 96, 667-700.

4. Guilmineau, E. (2008) Computational Study of Flow around a Simplified Car Body. Journal of Wind Engineering and Industrial Aerodynamics, 96, 1207-1217.

5. Regert, T. and Lajos, T. (2007) Description of Flow Field in the Wheelhouses of Cars. International Journal of Heat and Fluid Flow, 28, 616-629.

6. Launder, B.E. and Spalding, D.B. (1974) The Numerical Computation of Turbulent Flows. Computer Methods in Applied Mechanics and Engineering, 3, 269-289.

7. Fox, R.W., McDonald, A.T. and Pritchard, P.J. (1985) Introduction to Fluid Mechanics. Vol. 7, John Wiley & Sons, New York.

8. Gillespie, T.D. (1992) Fundamentals of Vehicle Dynamics. Vol. 400, Society of Automotive Engineers Warrendale.

9. Stern, F., Wilson, R.V., Coleman, H.W. and Paterson, E.G. (2001) Comprehensive Approach to Verification and Validation of CFD Simulations—Part 1: Methodology and Procedures. Journal of Fluids Engineering, 123, 793-802.

10. Wilson, R.V., Stern, F., Coleman, H.W. and Paterson, E.G. (2001) Comprehensive Approach to Verification and Validation of CFD Simulations—Part 2: Application for Rans Simulation of a Cargo/Container Ship. Journal of Fluids Engineering, 123, 803-810.

11. Duell, E.G. and George, A.R. (1999) Experimental Study of a Ground Vehicle Body Unsteady near Wake. Technical report, SAE Technical Paper.

12. Landström, C., Löfdahl, L. and Walker, T. (2009) Detailed Flow Studies in Close Proximity of Rotating Wheels on a Passenger Car. Technical Report, SAE Technical Paper.

13. Cengel, Y.A. and Cimbala, J.M. (2006) Fluid Mechanics. Vol. 1, Tata McGraw-Hill Education, New York.

14. White, F.M. and Corfield, I. (2006) Viscous Fluid Flow. Vol. 3, McGraw-Hill, New York.

CHAPTER 7

Electric Vehicles Analysis inside Electric Mobility Looking for Energy Efficient and Sustainable Metropolis

Miguel Edgar Morales Udaeta[*], Carolina Attas Chaud, André Luiz Veiga Gimenes, Luiz Claudio Ribeiro Galvao

Energy Group of the Electrical Power and Automation Engineering, Department of the Polytechnic, University of São Paulo—GEPEA/EPUSP, São Paulo, Brazil

ABSTRACT

This paper aims to study and evaluate electric mobility over time, focusing on the development of the electric car. Methodologically, in order to accomplish this intent, the characterization of the electric vehicle (EV) is made based on the variables which determine its performance, such as: assessment of speeds, distance traveled, analysis of facts related to the energy source (electro- chemical accumulators) and analysis of the determining system of electric mobility (the electric engine as a function of power (W) and voltage (V)). This way, to demonstrate the effects of time, this process will be analyzed from the beginning of the 20th century (1930s) to the present (the first decade of the 21st century), methodologically structured in 4 cycles that show the performance of the EV. The results show the existence of vulnerabilities and of electric mobility potential, as well as the nuances of the development of the electric vehicle along the years and along the transformations in what is considered state-of-the-art. Thus, in the case of batteries, it is evident that the lithium-ion type used nowadays reveals better results due to its higher specific efficient energy, which maximizes energy autonomy to 200 km. In the beginning, the insertion of the electric vehicle was commercially harmed by the fundamental limitations of batteries as a power source. Conclusively, on certain occasions there have been improvements in the aerodynamics, engines, weight and size of the batteries, demonstrating the maturity of EVs.

KEYWORDS

Electric Vehicle, EV Development, Electric Mobility, Energy Analysis, Metropolis, Sustainability

1. INTRODUCTION

Urban mobility in the 21st century comes with a series of constraints—ranging from problems in availability of essential resources to difficulties in the disposal of waste by average speed in the transportation of goods and people—all of which leading to considerations about the future role of the automobile in the cities. This is an opportunity to think and implement fundamental changes in concepts and models of the last century's standardized transport. Even though cars, trucks and other mobility machines have been synonymous with urbanization development, nowadays these machines are partially responsible for the collapse of metropolises [1] . From numerous negative factors relating to urban mobility in the full sense, the biggest target for the minimization of effects is local pollution, simply because humans need well-being. In this sense, electric mobility seems as a part of a solution known to the cities since—on the contrary of what it may seem—the first electric vehicles have emerged in the turn of the twentieth century, preceding the invention of the combustion engine gas (by Daimler Benz in 1885's Germany). In fact, before the internal combustion engine imposed itself as synonymous with mobility, the history of western socioeconomic development demonstrates that the electric vehicles industry was thriving, with electric buses lines, for example, which earned space on the streets of London around 1886. This has occurred because of the supremacy of the research conducted in France by Gustave Trouvé in 1881, which has made it possible to recharge batteries.

Regular use of electric vehicles has endured a long time, the proof of which being the remarkable advances in this area, such as the construction of the electric car that reached the incredible speed, for those days, of 100 km/h (by the Belgian Camille Jenatzy in 1899). The same way, in 1918, the electric buses line between Praça Maua and the then existing Monroe Palace in the city of Rio de Janeiro was inaugurated, known as the comfortable bus battery—powered electric traction—with solid rubber wheels, no noise, no vibration and without smo- ke or the hassle of gasoline (Peres, 2003).

The entrance of the Ford T in the market in 1909, along with its subsequent improvements such as the electric start, meant at the time the fall of the electric car—even with the advances recharge time and autonomy. Thus, like T Ford, urban mobility initiated its way to the current days. More than that, in the consumer society the automobile has become an indispensable asset. The following well-known step, in the case of the internal combustion engine, was the investment in oil refinery fields and their availability primarily in the U.S., which has determined the necessary support fuel. It is important to recall that the diesel engine [2] had been invented in the late 19th century but the use of fossil fuels like diesel in general transportation is not well seen due to the serious pollution and GHG emissions that result from it.

In this sense, environmental initiatives have favored in recent times the technological evolution of the electric power train, such as the Kyoto protocol, which especially concerns the burning of petroleum in powered internal combustion engine vehicles. Additionally, these vehicles in large population centers are a significant source of contaminating emissions, contributing often in 100% with the emission of pollutants into the atmosphere [3] [4] . Therefore, electrical and urban mobility are inserted in a new context due to the increase in the availability and use of electric vehicles in the world.

It is interesting to evidence that the relative emission of urban mobility—fueled by petroleum derivatives— while added to the combustion products of local industries becomes lethal when the case involves temperature inversions that prevent the dispersion of pollutants [5] .

Therefore, to foster the construction of an eco-economy towards a sustainable socioeconomic development that includes mobility in large cities seems irreversible in the 21st century. This means we live in a world where energy comes from clean sources such as wind turbines and not from coal mines, where recycling industries replace mining industries and where cities are planned for people and not for cars. A sustainable economy includes the welfare of future generations [1] .

As a premise, the electric vehicle proves itself to be conducive to environmental issues of urban mobility, since emissions are significantly reduced, which should also include a decrease in noise. Since recharging batteries (neuralgic point of the inclusion of electric mobility) incorporates new concepts into the commercial power grid, the inclusion of the electric vehicle should be planned within the power generation mix of each country. In any case, with respect to Brazil—given that the electric energy is hydroelectric—electric mobility is a significant option for sustainable urban mobility in megacities such as São Paulo [2] .

Relevant Aspects of the Electric Vehicles

At first, an electric vehicle in its basic meaning can be understood as being an automobile with an electric motor connected to the front wheels through a gearbox with one or two speeds, but there are several other possible variations of the propulsion system architectures. Thus, for example, a significant variation of this techonology is the use of four small motors for each wheel, in exchange for only one drive motor, as originally designed in the beginning of the electric car [6] .

Moreover, in the context of this work, electrical mobility aims at the driving of people, objects or a specific load. However, the electric vehicle in this case, independently from other technological variants, is generally understood as a system whose only power source is the battery charged with the task of activating one or more of the automobile's electric motors. Even because, in this specific case of electric mobility, it is generally assumed that the availability of power supply determines the type of vehicle so that, for example, a subway or trolley follows a predetermined route by rail or electrical distribution network [7] .

Given the diversity—in the modern world—of electric mobility among existing systems for electric vehicles, this work focuses on cars and utility vehicles for loads. In this sense, the study considers electric vehicles throughout time, in an analysis divided by the phases of its evolution, considering also the technology employed and socioeconomic importance of the period.

2. TECHNICAL EVOLUTIONARY CONTEXT OF ELECTRIC VEHICLES

In essence, the electric mobility's technology is based on the joint implementation of energy accumulators and of the electric motor as drive system. Two researchers stand out in this context: Alessandro Volta for the precursor of the battery in 1800 ("Battery Back") and Michael Faraday, who developed the homopolar motor in 1821.

In this sense, the predominance of the electromagnetic induction since 1831 has led to the consolidation of the electrical and electronics technologies, including engines and electrical generators. From this assumption, all findings related to the operation of an electric motor have supported the conjunction of the battery and of the electric motor connected to the wheels of a light vehicle. The first electric vehicles appear in the 1830s using non-rechargeable batteries.

In 1859 Gaston Plante introduced the rechargeable lead-acid, which is the technology currently used in acquiescence in most applications requiring energy storage. Following that, Aphonse Camille Faure improved the ability of such batteries, leading to industrial scale production, in such a way that in 1881 the autonomous electric vehicles proliferated the cities' streets. That way, it is important to mention that in the late 20th century, with the rechargeable batteries, electric cars entered the commercial market [8] . Thus, it can be said that the predominance of the accumulation of electrical energy and its conversion to mechanical energy has enabled a new, quiet and clean method for urban mobility.

As for the accumulators, other batteries have been developed—for example, the iron-zinc battery. Even Thomas Edison, interested in the potential of electric vehicles in the early 1900s, developed the nickel-iron battery, with a storage capacity 40% greater than that of the battery lead, but with a much higher cost of production. Thus, the 19th century saw the development of batteries such as the nickel-zinc and the zinc-air [9] .

Thus, electrical mobility starts when the first electric vehicle (tricycle) to use lead-acid (developed by Planté) as a source of energy was demonstrated in France by Mr. Trouvé in 1881. In this period, other electric tricycles with lead batteries were also in the U.S. and UK. In this context, it is worth remembering that only in 1885 did the German Karl Benz demonstrate the first combustion powered vehicle: the Pantentmotorwagen [10] .

Moreover, in 1837, Robert Davison Aberdeen built in England the first electric carriage, powered by a rustic iron-zinc battery and driven by an electric motor, but containing all the basic elements used in modern electric vehicles. In France, experiments were performed by Charles Jeantaud Raffard while, at the same time, Werner Siemens, in Germany, perfected the electric motor. Even though the mobility steam prevailed at the time—es- pecially in the area of public transport—the electric vehicle showed itself ideal for urban traffic, not having made noises and having had a drive system that did not pollute the environment [7] .

Anyway, since its inception electric mobility has gained significance, seeing that the electric motor, by not involving any combustion, was free of soot and grease, therefore being very clean. More than that, the electric vehicle with graduations for three, four and up to nine speeds was not required to carry a paraphernalia command, which is a fundamental characteristic of petrol and steam cars [11] .

3. FIRST CYCLE OF ELECTRIC VEHICLES (1837-1912)

Schiffer [12] points out that the electric car powered by rechargeable batteries seemed to have a great future about a century ago. Twenty-eight percent of the 4192 cars produced in the U.S. in 1900 were electric. Some of the most prestigious inventors, including Thomas Edison, promoted electric cars or the participation in its development. And the first industries to produce cars in series were manufacturing electric cars. In the early twentieth century, electric, steam and gasoline cars competed more or less on equal terms. Many analysts at the time believed that each type of car would find its own "performance space" and that they would coexist indefinitely. However, by the late 20s the electric car was a product in decline when considering the commercial sphere. The gasoline-powered car had conquered all spaces with its impressive speed, performance and finish. "Tripping A spectacular", from "The electric automobile in America". Table 1 presents schematically an analysis of the performance achieved each year by the respective electric vehicles, parameterized in speed, range, battery voltage and power, in the golden period related to the 1st cycle, which it set between 1837 and 1912.

Occasionally, around 1905 the gasoline-powered vehicle began to stand out when compared to the electric vehicles in question. The autonomy of the 63 miles (about 100 km) reached by vehicle combustion more than doubled the range of an electric car (30 miles or approximately 50 km), as seen in the table. The initial investment and the operating cost of electric cars were higher than the gasoline-powered. The available figures indicate that the 1900s petrol cars cost between $1000.00 and $2000.00, while an electric car cost U.S. $1250.00 to U.S. $3500.00. The cost of operating a gasoline car was $0.01/mile while for an electric car it was $0.02 to 0.03/mile. In 1901 big oil fields were discovered in Texas, which drove down petrol's costs in a way that it gained a sustainable competitive advantage.

Still, according to Larminie and Lowry [8] , early in the development of electric vehicles, an internal combustion engine that flipped a generator was used together with one or more electric motors, called hybrid vehicle.

According to Schiffer [12] , electricity best fulfills the requirements of a traction system than steam engines or even internal combustion engines.

However, in 1912, the fleet of electric cars in the United States reached its peak of 30,000 units and the amount of petrol cars was already thirty times larger, to the verge of 900,000 units [13] . Given this background, in the same period, between 1900 and 1912, there were initiatives in the pursuit of improving distance and performance of electric vehicles through the adoption of the hybrid configuration.

Based on the assumptions presented above, the trajectory of electric cars has continued in eversion. Among the main factors that have since contributed to the decline of electric cars [14] , the following can be cited:

- The competitive advantage achieved by the development of the production system in series, applied by Henry Ford, has allowed an increase in the manufacture of combustion-powered cars, as shown in Table 2.

- Elimination of the crank used to drive vehicles powered combustion. Invention of the electric starting, in 1912 [15] ;

- In the mid-1920s, the highways in the United States interconnected several cities, which demanded vehicles capable of traveling long distances [14] ;

- Oil discoveries in Texas have reduced the price of gasoline, making it an attractive fuel for the transportation sector [16] ;

- Development of distillation techniques in continuous and consequent cheapness of petroleum products resulting in the expansion of the technological development of the automobile industry headed for gasoline-powered vehicles [7] .

Table 1. Evolution of electric vehicles from 1837 to 1912.

Year	Speed	Autonomy	Battery	Motor (Power/Voltage Type)		
1837	6 km/h	2 km	Zinc-acid	5 KW	-	-
1881	15 km/h	40 km/h	Lead	0.37 KW	20 V	MCC
1890	14 km/h	23 km	24 cells	3 KW	58 V	-
1897	15 km/h	48 km	Lead	2.6 KW	-	-
1899	105.8 km/h	-	Lead	2 × 50 KW	200 V	MCC
1900	58 km/h	-	Lead	2 × 5.15 KW	80 V	-
1902	21 km/h	64 km	Ni-Fe	1 KW	40 V	-
1908	16 km/h	-	40 cells	-	40 V	-
1911	37 km/h	60 km	Ni-Fe	6 KW	84 V	-

Source: own compilation.

Table 2. Number of vehicles built in USA in the beginning of the century.

Year	Electric	Gasoline
1899	1575	936
1904	1425	18,699
1909	3826	120,393
1914	4669	564,385
1919	2498	1,649,127
1924	391	3,185,490
1929	757	4,454,421
1933	-	1,560,599

Source: own adaptation from [?].

4. SECOND CYCLE OF ELECTRIC VEHICLES (1912-1973)

After the discovery of oil fields, gasoline vehicles and, later, diesel quickly reached levels of performance that resulted in greater speed, greater acceleration and lower weight compared to electric vehicles. The oil industry had developed into such a supreme point that virtually all of its derivatives started to present cost advantage because of its growing consumption.

Sales were minimal even in England, where Brougham (by Partridge Wilson) was marketed, powered by a battery of 60 V and 34 Ah, which enabled a speed of 51.5 km/h and a radius of 97 km per battery charge [7] .

Based on the assumptions presented above, having the 2nd cycle started around 1909, the production of electric cars dropped considerably, reaching about 4.4% of the number of combustion-powered cars. In 1913, as aforementioned, Ford began producing gasoline cars in series in the first industrial assembly line in Highland Park plant. In 1912 with the advent of the electric starter motor vehicles, the explosion made these cars even more attractive.

Obviously, by 1912 there was renewed enthusiasm for the electric car with the emergence of a few technical developments. Thomas Edison had perfected and carried out the first tests using the nickel-iron battery, which had a 35% increase in storage capacity between 1910 and 1925. The lifespan of these batteries has also increased whereas the maintenance costs decreased.

However, this resurgence was most striking for small delivery trucks in companies that owned fleets of around 60 vehicles and that could have their own central recharge batteries.

The advent of the First World War in 1914 caused an increase in oil prices and also in optimism about electric cars. But despite the commercial and marketing efforts, the number of electric trucks fell from 10% in 1913 to only 3% - 4% in 1925 according to Schiffer [12] .

However, the years between 1920 and 1970 were a time of steady decline in electric cars. On a global scale, the depression of the 1930s, followed by World War II, harmed a possible resurgence of electric vehicles and new experiments with alternative fuel vehicles. At that time, few studies and scientific research have been developed for electric vehicles. Even in the postwar period of economic prosperity, the electric vehicle projects remained stored in a context of small concern for energy security, due to the existence of abundant and cheap fuel and vehicles with internal combustion engine (MCI), the largest and fastest [15] .

In the United States, a meager revival of electric vehicles happened in the 60s. Car technology was basic, with DC motors with brushes. However, in terms of energy accumulators there was some variety beyond the normal lead-acid: the lead-cobalt and the nickel-cadmium.

Specifically, the electric traction technology began again to be shyly explored, its development returning only from the 60s on, when the electric vehicle was seen as a way to overcome the environmental problems caused by emissions from the combustion powered vehicles. It is noteworthy that most of the electric vehicles produced in the 60s emanated from the conversion of conventional vehicles.

We observe, however, that from the 1970s on, with the emergence and worsening of the oil crisis, discussions on environmental issues in urban centers have become a worrying factor for government leaders. The electric vehicle has been considered as an energy alternative, mainly in countries with a lot of hydroelectric generation or coal-based thermal power. In this period, there were several initiatives aiming to insert them back into the market, but neither the pure electric cars nor the hybrids were able to compete in the market with conventional cars, which had a sustainable competitive advantage [9] .

Thus, the performance achieved each year by the respective electric vehicles parameterized in speed, range, battery voltage and power on the 2nd cycle (between 1912 and 1973) are summarized in Table 3.

Because of its importance and relevance to this study, as a kind of complementary measure, the table shows one of the striking points in time due to short range (50 to 100 km) and low average speed (on average, 50 to 100 km/h). Somehow, these factors affected the introduction of electric vehicles on a large scale. Moreover, the delay on the recharging of the battery (about 8 hours) and the lack of infrastructure in use service became a concomitant factor.

Meanwhile, according to Valle Real and Balassiano [17] , there are basically two paths to be taken. The first would make them more efficient vehicles from the point of view of energy consumption (as well as of the amount of emissions); and the second, by means of restrictions and adoption of specific rates, would make the user reduce the utilization of motor vehicles, particularly the car.

5. THIRD CYCLE OF ELECTRIC VEHICLES (1973-1996)

Apart from being cleaner (depending on the type of energy used), electric cars manufactured and converted in the late 60s used certain conservation techniques in an attempt to increase their autonomy and maximum speed. The priority was in guaranteeing that these vehicles reached the level of performance offered by combustion powered cars, whose development had been significant throughout the century.

Interest in electric vehicles had increased considerably until the late 80s, when the problem of air pollution from large cities began to be discussed more often [6] . With respect to Brazil, for example, the fundamental importance of engineer John Augustus Conrado do Amaral Gurgel (1926-2009) cannot be forgotten, since he produced the first Brazilian electric car in 1974, with a range of 60 km—called Itaipu.

Generically speaking for this period, it is noteworthy that by the year 1973 the crisis associated with the embargo imposed by the OPEC (Organization of Petroleum Exporting Countries) brought new prospects for electric cars. However, the United States depended significantly on oil from the Arab countries and the U.S. Congress at the time was determined to reduce this dependence. There was also an economic motivation given by the U.S. trade balance. Environmental issues were not actually critical, therefore, it was not generally considered that only the use of electric cars would improve air quality, according to Schiffer [12] .

Objectively, as stated before about the 80s, the government interests turned to the advantages derived from powered electric vehicle propulsion vehicles, mainly because of environmental issues. Thus, it is observed that government policy measures were introduced worldwide in the pursuit of reducing urban vehicle emissions. The prime example was the California Air Resources Board's

(CARB) that implemented, in 1990, the first regulatory standards for zero-emission vehicles in California.

Table 3. Evolution of electric vehicles from 1912 to 1973.

Year	Speed	Autonomy	Battery	Motor (Power/Voltage Type)		
1915	50 km/h	161 km	Lead	32 KW	76 V	-
1916	42 km/h	60 km	Lead	18 KW	80 V	-
1917	40 km/h	340 km	Ni-Fe	-	-	-
1941	60 km/h	50 km	Lead	-	96 V	-
1947	75 km/h	65 km	Lead	-	36 V	-
1960	70 km/h	160 km	Lead	2 × 6 KW	48 V	-
1961	50 km/h	55 km	Lead	2 × 6 KW	48 V	MCC
1966	100 km/h	210 km	Pb-Co	57 KW	120V	MCC
1967	60 km/h	60 km	Lead	3.7 KW	48 V	-
1968	85 km/h	190 km	Ni-Cd	-	-	MCC
1972	60 km/h	140 km	Lead	32 KW	144 V	-
1973	85 km/h	60 km	Lead	3 × 2.5 KW	48 V	-

Source: own compilation

A number of modern vehicles was introduced by automakers between the 1980s and 1990. Along with the selling of the top vehicles, such as the General Motors EV-1, the Toyota RAV4-EV and the Ford Ranger EV, several studies about the cost of batteries have been developed in order to assess the commercial prospects of these vehicles [12] . Although they were more efficient than conventional cars, this advantage had little value at the time when the oil price was the lowest in history [18] .

Based on the performances achieved each year by its respective electric vehicles, parameterized in speed, range, battery voltage and power and in comparison to previous cycles, Table 4 shows the variables from the 3rd cycle, between 1973 and 1996.

6. FOURTH CYCLE OF ELECTRIC VEHICLES (1997 TO THE PRESENT)

Within the different stages of live cycles, currently the hybrid, the electric and the plug-in vehicles emerge as instruments to solve flagship-oriented issues such as energy security and climate impact. According to Anfavea [19] , the Brazilian Chain of Automotive Supplies brings together a spectrum of ethnic diversity in the nationalities of their manufacturers, so that automakers are gathered here from no less than nine different countries: Germany, Brazil, South Korea, United States France, India, Italy, Japan and Sweden. This ethnic diversity has no record in any other major producer of vehicles on the planet [19] .

Table 4. Evolution of electric vehicles from 1973 to 1996.

Year	Speed	Autonomy	Battery	Motor (Power/Voltage Type)		
1974	60 km/h	70 km	Lead	4.4 KW	48 V	MCC
1976	60 km/h	90 km	Lead	8.8 KW	72 V	MCC
1977	105 km/h	100 km	Lead	17 KW	84 V	MCC
1978	120 km/h	160 km	Lead	24 KW	36 V	-
1980	105 km/h	115 km	Lead	18 KW	96 V	MCC
1981	81 km/h	80 km	Lead	15 KW	102 V	-
1983	80 km/h	110 km	Lead	10 KW	72 V	-
1984	50 km/h	115 km	Lead	24 KW	84 V	-
1987	100 km/h	80 km	Na-S	17 KW	200 V	MCC
1989	90 km/h	100 km	Lead	18 KW	96 V	MCC
1990	105 km/h	150 km	Lead	60 KW	320 V	MI
1991	120 km/h	276 km	Ni-Cd	30 KW	200 V	MCC
1992	110 km/h	170 km	Na-NiCl	62 KW	120 V	MCC
1993	120 km/h	90 km	ZnBr2	12 KW	168 V	MI
1994	85 km/h	100 km	NiCd	30 KW	240 V	MI
1995	90 km/h	160 km	Lead	2 × 18 KW	72 V	MCC
1996	100 km /h	90 km	Li-ion	30 KW	216 V	MI

Source: own compilation.

Most of these automakers that help compose the chain of the automotive segment conduct research with universities in order to develop integrated models of hybrid and electric cars format. The first major step in this recent movement arose in 1997, when Toyota, the Japanese automaker, launched the Prius in Japan—a hybrid four-door sedan—followed by Honda, the first to launch a hybrid in the U.S. market—the Insight in 1998 [20] .

Since the launch of the Toyota Prius in 1997, 1.9 million HEVs (Hybrid Electric Vehicles) vehicles and 60.0 thousand PHEVs (Plug-in Hybrid Electric Vehicles) and BEVs (Battery Electric Vehicles) vehicles have been sold in the North-American market [15] . This fact can be attributed in large part to the encouragement of the U.S. government to manufacturers and consumers of hybrid and electric vehicles. Worldwide, over the last decade, many HEVs,

PHEVs and BEVs have been sold, totaling more than 2.5 million vehicles. In early 2011, the penetration of these technologies in the market is of 2% in the U.S. and 9% were sold in Japan [21] . A descriptive approach to the start of the 4th cycle will be presented below.

Although it is necessary to prioritize the previously mentioned variables that portray each historical cycle, at first, it is noticed that the electric motor is the ideal drive propulsion. Due to the strong competition and the growing consumer demand, companies in the area of transport and logistics attempt to reduce operating costs, while seeking to improve services. The electric vehicle meets those needs, since it has compatible attributes such as: it is quiet, it is highly efficient, it has excellent torque characteristics × speed and it does not pollute. However, the Mercedes-Benz CL600 has a 367 hp of 12-cylinder engine and can accelerate its 2380 kg 0 - 60 miles/h (about 100 km/h) in 6.3 s. To increase efficiency there is a mechanism that disables 6 cylinders when there is no need for high torque, according to Dettmer [22] . Table 5 shows the variables portrayed in the 4th cycle, starting in 1997 until today.

With the new developments in batteries, electric vehicles now have their storage capacity between 20 and 60 kWh, allowing its interconnection with the electric distribution network through the consumption of energy. Plus, in the very near future, this will provide energy according to the needs of network functionality through the Vehicle Connected to the Network (VLR), according to Kempton and Tomic [23] .

Figure 1 shows a partial configuration of the components of an electric vehicle that properly uses the electricity from the public distribution system to recharge the battery installed in the vehicle (battery bank). The received energy is stored in the battery, in electrochemical format. This stored energy is converted into electrical energy that is transported to the Electric Motor (M/G), which will make its conversion into mechanical energy and thus provide the movement of the vehicle without generating emissions or noise. Also, if a Regenerative Breaking System (TR) is implemented in the electric vehicle, it is also possible to store the energy produced when breaking or slowing down, by converting the kinetic energy into electrical energy through the M/G, which will be stored in the battery [24] .

7. POTENTIALITIES AND LIMITATIONS HIGHLIGHTED IN THE LV DEVELOPMENT

The importance of batteries is irrefutable in an electric vehicle. That way, the main characteristics of an accumulator of energy are the specific energy, specific power and lifetime. Specific energy is the energy amount stored by the battery per unit weight; the specific power is the supplied power per unit mass; and the lifetime is the number of charge/discharge that may be required.

Table 5. Evolution of EVs between 1996 and the present.

Year	Speed	Autonomy	Battery	Motor (Power/Voltage Type)		
1997	100 km/h	100 km	NiMH	18.5 KW	288 V	MS
1998	120 km/h	185 km	NiMH	84 KW	-	MI
1999	100 km	115 km	Li-ion	24 KW	300 V	MIP
2000	130 km/h	200 km	Li-ion	65 KW	345 V	MIP
2002	120 km/h	80 km	NiCd	44 KW	180 V	MIP
2007	150 km/h	290 km	Li-ion	150 KW	355 V	MI
2008	110 km/h	100 km	Li-ion	2 × 15 KW	144 V	MIP
2009	150 km/h	210 km	Li-ion	150 KW	380 V	MI
2010	140 km/h	300 km	Li-ion	200 KW	380 V	MIP
2011	145 km/h	270 km	Li-ion	47 KW	360 V	MI
2012	120 km/h	120 km	Li-ion	55 KW	300 V	MIP

Source: own compilation

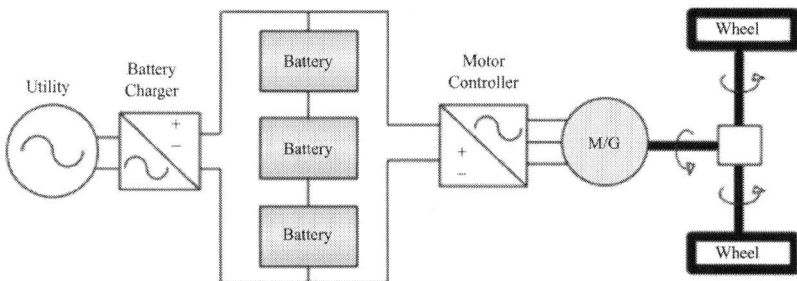

Figure 1. Evolution of batteries.

In Graph 1, it is possible to check, on a timeline, the progress in size and weight of the various types of batteries. The batteries of the Pb-SO$_4$, Ni-Cd, Ni-MH type are stagnant in terms of specific energy and density.

Thus, it can be defined as energy density the ratio between the maximum amount of energy that can be stored securely in an energy storer element body and the volume of that body. The higher the energy density—which can be measured in Wh/l (watt/h per liter) or MJ/l (megajoules per liter)—the more energy can be stored or transported in one body with the same volume. As for the specific energy, it also relates to the maximum amount of energy that can be stored, but with the mass of the containing body element. Specific energy can be quantified in Wh/kg (watt-hours per kilogram).

Peças Lopes, Soares, Almeida and Moreira da Silva [11] point out that, with the new developments of batteries, vehicles powered by electric traction have the storage capacity of between 20 and 60 kWh, allowing its interconnection with the power grid distribution through the energy consumption. In the very near future, this will provide power according to the needs of the network through the functionality of a Vehicle Connected to the Network (VLR).

On the face of it, the technical requirements demanded by the energy accumulators are different for each type of vehicle. Electric vehicles require batteries with higher energy densities, which limits them due to the masses and volumes associated, contributing to the low range of these types of vehicles. Since in pure electric vehicles batteries are the only energy source, these end up suffering deeper discharges, demanding more robust batteries with long life and the acceptance of a high number of charge and recharge cycles. However, increasing the autonomy of electric vehicles requires larger batteries, greatly increasing the vehicle mass. Inversely, reducing the range of electric vehicles enables higher effective energy efficiency. In Graphs 2-5 we can see the evolution of the range and speed before 4 cycles, taking into account previously shown factors that were considered essential for the development of these vehicles.

It is worth noting that the lead-acid (Pb) batteries are the best-known being used in cycles 1, 2 and 3, having the largest implementation in 2012. These batteries are the most inexpensive and they require little maintenance; however, they have limited power and specific energy, 40 Wh/kg and 350 W/kg respectively. The average lifespan of these batteries is one of its limitations, being it about 500 charge/discharge cycles.

Faia [25] discusses that the type of battery which has appeared most promising in recent years has been the lithium (Li-ion) ion. This battery, used intrinsically in the 4th cycle, has a specific energy of over 150 Wh/kg and specific powers that can go up to 2000 W/kg. Its lifespan is about 1200 charge/discharge cycles. The disadvantages of this type derive from the fact that batteries require a precise charging system due to its low tolerance for peak power and are still relatively expensive for pure electric vehicles.

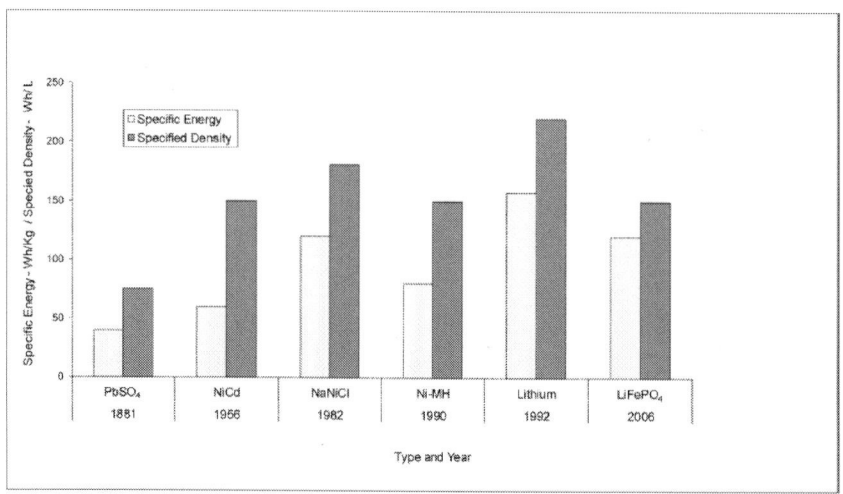

Graph 1. Evolution of batteries.

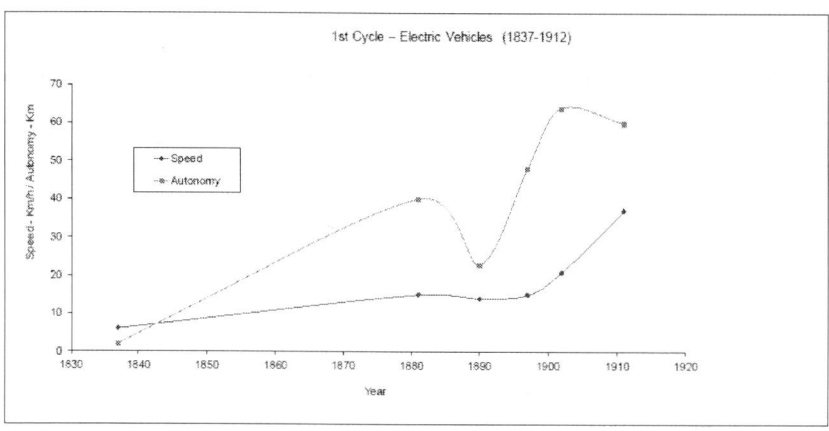

Graph 2. 1st cycle—electric vehicles (1837-1912).

A solution to the inflexibility of the batteries' autonomy has been presented by Andersen et al. [26] , being implemented in Israel. The core of this proposal is the separation of the properties of the car and of the battery. A company would be responsible for assuming the risk of ownership of the batteries, which would be leased. The consumer would be charged for the energy they consume over the kilometers traveled. This is analogous to the collection of minutes used by mobile phone operators. Consumers would have a number of "packages" available, which will depend on the battery usage profile.

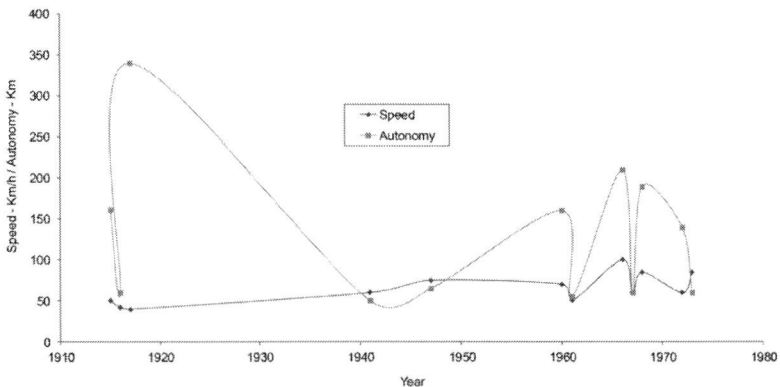

Graph 3. 2nd cycle—electric vehicles (1912-1973).

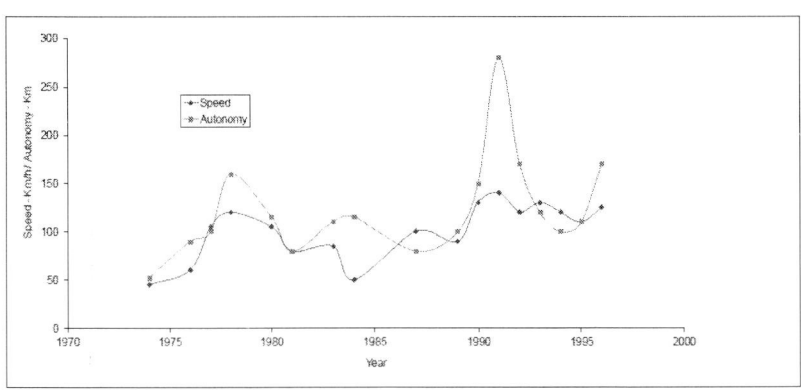

Graph 4. 3rd cycle—electric vehicles (1973-1996).

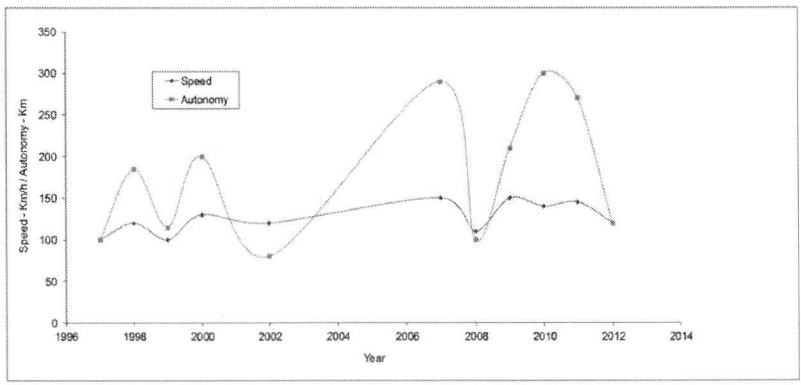

Graph 5.4th cycle—electric vehicles (1996 to present).

In this aspect, the main types of motors used in electric vehicles are: Continuous Current Motor (MCC), or Asynchronous Induction Motor (IM), Permanent Magnet Synchronous Motor (IPM) and Switched Reluctance Motor (MRC). In Figure 2 one can see the qualitative assessment of various engine characteristics, with the objective of identifying the technologies that would have the best performance in the electric car applications according to Zeraoulia, Benbouzid and Diallo [27] .

It is possible to compare the total ratings of the characteristics of the different propulsion systems listed in Figure 2, allowing us to infer that the MI OE MIP are the most suitable for electric vehicles. Thus, from the 4th cycle to the IPM motor cycle, this type of vehicle has currently become more solicited because of its gains in density, its efficiency and its cost.

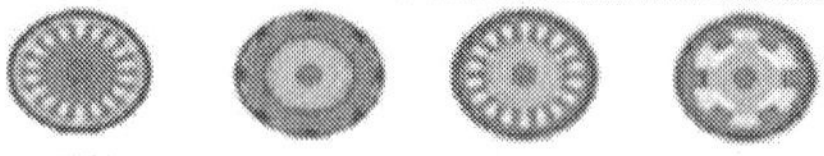

Characteristics	MCC	MI	MIP	MRC
Voltage Density	2.5	3.5	5	3.5
Density	2.5	3.5	5	3.5
Controllability	5	5	4	3
Reability	3	5	4	5
Technological maturity	5	5	4	4
Cost	4	5	3	4
∑ Total	22	27	25	23

5   - Excellent
4   - Very Good
3.5 - Good
3   - Sufficient
2.5 - Less than sufficient

Figure 2. Matters of electric motor selection drive for HEV propulsion systems: a comparative study.

8. CONCLUSIONS AND FINAL THOUGHTS

The analysis in the context of this study points to the vulnerabilities and potentialities that derive from the introduction of electric vehicles (EV) in a timeline. In essence, the lithium used properly in the fourth cycle ion shows better results due to its higher specific energy and high energy efficiency maximizing autonomy of 200 km. Moreover, there have been improvements in aerodynamics, engines, and the weight and size of batteries, demonstrating the technical and marketing maturity of the EVs. That is why studies about the logistics of such vehicles in terms of autonomy become intrinsically relevant.

Accordingly, for the first two cycles, specifically in the context of a car, all variables described by means of Tables 1-3 show that the inclusion of these commercially electric vehicles is hindered by the fundamental limitations of batteries as a power source.

Also in relation to energy accumulators, known as batteries, it is valid to emphasize the significant time (6 - 12 pm) they take to be recharged. In contrast, a gasoline tank takes about 2 - 3 minutes to be completed, which means a power flow of an order of 20 - 30 MW during its supply in a gas station—an issue addressed by Hermance and Sasaki [28] .

Undoubtedly, gasoline has become an ideal fuel, but it also has its adversity such as very low efficiency and, in low rpm conditions, the available torque is low. Torque is what determines the acceleration capability; for a conventional car, the combination of the gearbox and the oversizing of the combustion engine—which carries an even greater inefficiency—is what defines this acceleration.

Once again, the main changes that took place, especially in the 3rd and 4th historical cycles of electric vehicles, should be emphasized. They range from the break-up of new technologies to the development of new motors, power converters, battery chargers and energy accumulators.

It is evident that the specific properties of electric vehicles vary according to the battery size and type of vehicle. The same way electric cars need higher specific energy per volume, hybrid vehicles in return require batteries that provide maximum power in the smallest possible size [25] .

In a simplified way, after the energy crisis of 1970 and 1980, and with the significant increase in emissions of Greenhouse Gases (GHG), the revival of the interest in electric vehicles based on energy sources alternative to oil arose, as is corroborated by Tomic and Kempton [23] .

In this sense, the reason why the production of electric cars remains static is related to some indicators such as high production costs—which translate into low penetration—, lack of logistics networks for the vehicles to refuel and unsatisfactory autonomy. This is undoubtedly a worrying factor for the countries, manufacturers and end customers throughout the automotive supply chain.

To firm the findings and considerations of this work, it is important to highlight the studies done by Hall, Reis and Junior [29] -[31] , which have proved that the lithium ion is expected to occupy a prominent place in electric vehicles in the present days. This is due to its higher specific energy in terms of volume and mass, its high energy efficiency (near 100%), its long life cycle (approximately

3000 cycles with a discharge depth of 80%), its low rate of self-discharge, and its lack of memory effect, which decreases battery capacity when it is recharged. Mass production and development of nanostructured materials provide considerable scope for cost savings. We observed their use in large scale from the year of 1996, during the fourth and last cycle analyzed.

Based on the factors put in evidence throughout this work, we point out that the disadvantages of electric vehicles are associated with deficiencies of electrochemical energy storage. Compared to conventional fuels, the battery of electric cars has low specific energy in terms of volume and mass and low rate of refueling/recharging. These questions are well evaluated by Bradley e Frank [32].

Additional considerations need to be brought up with respect to another aspect that should receive attention in EVs. One of them is associated with the use of electricity by the transportation sector. The introduction in scale of EVs comes with an increase in electricity demand and the possible need for increased capacity to generate electricity. More than that, the penetration of electric vehicles in the electrical system can cause overload on transformers and distribution lines, as well as—mainly depending on the timing and form of clearance—over- shoot in the electrical system. These aspects of the electrical system are also put in evidence by Kiviluoma and Meibom, Hadley and Tsvetkova, Green II et al., Lin et al. [33] -[36] , among others.

Finally, it is concluded from the results, based on the four cycles presented here, that the inverters have improved over the years. We highlight that in old electric vehicles' DC, there was the need for a transmission to enable reverse driving (rear gear), but this is no longer needed.

REFERENCES

1. Brown, L.R. (2003) Eco-Economy: Building and Economy to Earth. Publicado no Brasil pela Universidade Livre da Mata Atlântica (UMA), Primeira Edição, 368 p.

2. Pecorelli Peres, L.A. (2003) Electric Vehicles: Environmental and Energy Benefits. Rio de Janeiro and Cultural Research Noel Rosa.

3. Ministry of the Environment (MMA) (1999) Inspection Program Evaluation and Maintenance of Vehicles in Use in Rio de Janeiro. Document Prepared by the partnership LIMA (Interdisciplinary Laboratory Ministry of the Environment (MMA). COPPE/UFRJ as Part of the Project "Air Quality Management in Major Metropolises Brasileiras" under the Third Amendment to the Agreement MMA/Foundation COPPETEC 1999-CV-000054.

4. Campi, T.M., Rutkowski, E. and Lima Jr., O.F. (2004) Sustainability of Transportation Techniques. UNICAMP, Campinas.

5. Braga, A., Pereira, L.A.A. and Saldiva, P.H.N. (2002) Pollution and Its Effects on Human Health. Seminar on Sustainability in Energy Generation and Use, UNICAMP, Campinas, 20 p. http://www.bibliotecadigital.unicamp.br/

6. Delucchi, M.A. and Lipman, T.E. (2010) Lifetime Cost of Battery, Fuel-Cell, and Plug-In Hybrid Electric Vehicles, Chapter 2. In: Pistoia, G. and Elsevier, B.V., Eds., Electric and Hybrid Vehicles: Power Sources, Models, Sustainability, Infrastructure and the Market, Elsevier, Amsterdam, 19-60.

7. Bottura, C.P. and Barreto, G. (1989) Electric Vehicles. UNICAMP, Campinas.

8. Larminie, J. and Lowry, J. (2003) Electric Vehicle Technology Explained. John Wiley and Sons Ltd., Chichester.

9. Baran, R. and Legey, L.F.L. (2011) Electric Vehicles: History and Prospects in Brazil. BNDES, 33, 207-224.

10. Hoyer, K.G. (2008) The History of Alternative Fuels in Transportation: The Case of Electric and Hybrid Cars. Utilities Policy, 16, 63-71.

11. Peças Lopes, J.A., Soares, F.J., Almeida, P.M. and Moreira da Silva, M. (2009) Smart Charging Strategies for Electric Vehicles: Enhancing Grid Performance and Maximizing the Use of Variable Renewable Energy Resources. EVS24, Stavanger.

12. Schiffer, M.B. (2010) Taking Charge—The Electric Automobile in America. Smithsonian Institution Press, Washing- ton DC.

13. Struben, J.R. and Sterman, J. (2006) Transition Challenges for Alternative Fuel Vehicle and Transportation Systems. MIT Sloan Research Paper.http://ssrn.com/abstract=881800

14. DOE (2012) Department of Energy, Energy Efficiency and Renewable Energy. History of Electric Vehicles.http://www1.eere.energy. gov/vehiclesandfuels/avta/light_duty/fsev/fsev_history.html

15. Anderson, C.D. and Anderson, J. (2010) Electric and Hybrid Cars: A History. 2nd Edition, McFarland & Company, Inc., Jefferson.

16. Yergin, D. (1991) The Prize: The Epic Quest for Oil, Money, and Power. Free Press, Nova Iorque.

17. Valle Real, M. and Balassiano, R. (2002) Identify Priorities for Aadoção Mobility Management Strategies: The Case of Dobrio City January. Proceedings of the 10th Congress of Research and Training in Transportation, ANPET, Natal.http://www.ivig.coppe. ufrj.br/ ivig/ Paginas/teses-dissertacoes-artigos.aspx

18. Faia, S.M.R. (2006) Optimization of Vehicle Propulsion Systems for Fleet. Thesis of M.Sc., Instituto Superior Técnico, Lisboa.

19. ANFAVEA (2011) National Association of Automobile Manufacturers, Statistical Yearbook of the Brazilian Automotive Industry.

20. Dijk, M. and Yarime, M. (2010) The Emergence of Hybrid-Electric Cars: Innovation Path Creation through Co-Evo- lution of Supply and Demand. Technological Forecasting and Social Change, 77, 1371-1390.

21. IEA (2011) International Energy Agency, Technology Roadmap: Electric and Plug-In Hybrid Electric Vehicles.

22. Dettmer, R. (2001) Hybrid Vigour. IEE Review, 47, 25-28.

23. Tomic, J. and Kempton, W. (2007) Using Fleets of Electric-Drive Vehicles for Grid Support. Journal of Power Sources, 168, 459-468.

24. Kramer, B., Chakraborty, S. and Kroposki, B. (2008) A Review of Plug-In Vehicles and Vehicle-to-Grid Capability. National Renewable Energy Laboratory, 1617 Cole Blvd., Golden, CO 80401, USA.BEV-HEV-PHEV-FCEV.

25. Broussely, M. (2010) Chapter 13—Battery Requirements for HEVs, PHEVs, and EVs: An Overview. In: Pistoia, G. and Elsevier B.V., Eds., Electric and Hybrid Vehicles: Power Sources, Models, Sustainability, Infrastructure and the Market, Elsevier, Amsterdam, 305-345.

26. Andersen, P.H., Mathews, J.A. and Rask, M. (2009) Integrating Private Transport into Renewable Energy Policy: The Strategy of Creating Intelligent Recharging Grids for Electric Vehicles. Energy Policy, 37, 2481-2486.

27. Zeraoulia, M., Benbouzid, M.E.H. and Diallo, D. (2010) Electric Motor Drive Selection Issues for HEV Propulsion Systems: A Comparative Study.http://hal.inria.fr/docs/00/53/33/62/PDF/IEEE_VPPC_2005_ZERAO ULIA.pdf

28. Hermance, D. and Sasaki, S. (1998) Hybrid Electric Vehicles Take to the Streets. IEEE Spectrum, 35, 48-52.

29. Hall, P.J. (2008) Energy Storage: The Route to Liberation from the Fossil Fuel Economy? Energy Policy, 36, 4363- 4367.

30. Reis, N.A.O. (2008) The Hybrid Car as a Supplier-Consumer Element of Electricity-Battery Modeling. M.Sc. Thesis, IST University, Lisbon.

31. Junior, A.R.P. (2002) Regulation of Energy Demand in a Propulsion System for a Hybrid Electric Vehicle Series. M.Sc. Thesis, UFRGN, Natal.

32. Bradley, T.H. and Frank, A.A. (2009) Design, Demonstrations and Sustainability Impact Assessments for Plug-In Hybrid Electric Vehicles. Renewable and Sustainable Energy Reviews, 13, 115-128.

33. Kiviluoma, J. and Meibom, P. (2011) Methodology for Modelling Plug-In Electric Vehicles in the Power System and Cost Estimates for a System with Either Smart or Dumb Electric Vehicles. Energy, 36, 1758-1767.

34. Hadley, W.S. and Tsvetkova, A. (2008) Potential Impacts of Plug-In Hybrid Electric Vehicles on Regional Power Ge- neration. UT-Battelle, Oak Ridge National Laboratory, Oak Ridge.

35. Green II, R.C., Wang, L. and Alam, M. (2011) The Impact of Plug-In Hybrid Electric Vehicles on Distribution Networks: A Review and Outlook. Renewable and Sustainable Energy Reviews, 15, 544-553.

36. Lin, S., He, Z., Zang, T. and Qian, Q. (2010) Impact of Plug-In Hybrid Electric Vehicles on Distribution Systems. International Conference on Power System Technology (POWERCON), Hangzhou, 24-28 October 2010, 1-5.

CHAPTER 8

Numerical Validation of a Vortex Model against Experimental Data on a Straight-Bladed Vertical Axis Wind Turbine

Eduard Dyachuk * and Anders Goude

Division of Electricity, Department of Engineering Sciences, Uppsala University, Box 534, Uppsala 751 21, Sweden

ABSTRACT

Cyclic blade motion during operation of vertical axis wind turbines (VAWTs) imposes challenges on the simulations models of the aerodynamics of VAWTs. A two-dimensional vortex model is validated against the new experimental data on a 12-kW straight-bladed VAWT, which is operated at an open site. The results on the normal force on one blade are analyzed. The model is assessed against the measured data in the wide range of tip speed ratios: from 1.8 to 4.6. The predicted results within one revolution have a similar shape and magnitude as the measured data, though the model does not reproduce every detail of the experimental data. The present model can be used when dimensioning the turbine for maximum loads.

KEYWORDS

wind turbine; vertical axis turbine; force; measurement; open site; simulation; vortex model; dynamic stall

1. INTRODUCTION

The interest in vertical axis wind turbines (VAWTs) as an alternative to the conventional horizontal axis wind turbines (HAWTs) has been growing in recent years [1]. This is due to the potential of VAWTs to decrease the cost of wind energy [2,3]. VAWTs have several advantages over HAWTs: the generator of a VAWT can be located at the ground level, therefore excluding the concerns over the mass and size of the generator. The advantage of the lower center of mass (compared to HAWTs) is very important for floating platforms [4]. Additionally, the yawing mechanism is excluded for VAWTs, since they are omni-directional. Thus, the simplicity of the concept with only a few moving parts is one of the main advantages of VAWTs over HAWTs.

The complex aerodynamics of VAWTs imposes significant challenges on the simulation models. The flow velocity at the blades of the VAWT changes constantly during the turbine rotation, which causes the angle of attack to change during every revolution. The magnitude of the variation of the angle of attack increases with the decreased turbine tip speed ratio (TSR). At low TSRs, the blades of the VAWT experience the event of dynamic stall, which is associated with the rapid decrease in the lift and the increase in the drag force, reducing the torque on the turbine. At high TSRs, the flow velocity when passing through the turbine is decreased more than at low TSRs, and therefore, the flow expansion is prevailing at high TSRs.

The simulation models for VAWTs can be divided into three groups. The first group includes the finite element method (FEM) or the finite volume method (FVM), which are used to solve the Navier–Stokes equations within the commonly-available software for computational fluid dynamics (CFD). The second method is based on the vorticity equation, and the models are usually referred to as vortex models. The third method is based on the momentum conservation principle, and one of the most common and advanced momentum models is the double multiple streamtube model. The overview of the aerodynamic models for VAWTs can be found in [5,6,7]. A two-dimensional (2D) vortex model combined with the Leishman–Beddoes-type dynamic stall model is used in this study. The model combines the vorticity equation with experimental data, which results in the high computational speed of the model. This model gives the flow velocity field and is time dependent.

There is a lack of experimental data on the blade forces during one revolution for VAWTs. A series of the experiments were conducted by the Sandia National Laboratories in the 1980s, where VAWTs with curved blades (Darrieus turbines) were operated at open sites [8,9,10]. Other measured data concern small vertical axis turbines operating in wind tunnels or towing tanks with low operational

Reynolds numbers [11,12]. Since the force coefficients are dependent on the Reynolds number, the aerodynamics of large turbines are different from the aerodynamics of turbines operated in wind tunnels or towing tanks. Thus, due to high Reynolds numbers, the measured data from the Sandia National Laboratories are still used for the validation of simulation models [13,14,15,16].

This study assesses measurements from 2014 on a straight-bladed VAWT, which operates at an open site with the average Reynolds number of 300,000 [17,18]. A study on the power coefficient (CP) of this VAWT from 2011 has shown that the turbine reaches its maximum CP of 0.29 at the TSR of 3.3 [19]. However, the turbine diameter has increased from 6 to 6.5 m after mounting the load cell assembly, and the power coefficient is expected to be slightly different for the modified turbine as the turbine solidity has decreased. New experimental data on the normal forces on this VAWT are presented. The goal of the study is to describe the simulation model and to validate it against the experimental data. The normal forces are compared for the range of TSRs from 1.8 to 4.6, covering the dynamic stall region and the region of high flow expansion. The results and the capability of the model are analyzed.

2. EXPERIMENTAL DATA

The measurement data used in this study are obtained from the 12-kW VAWT located outside Uppsala, Sweden. The experimental method and the obtained force data are described in detail in [17,18]. It follows that the measured normal force was periodic and consistent, while the tangential force response was highly disturbed by the turbine dynamics. Hence, only the normal force measurements can be considered suitable for usage in this validation. The studied VAWT is a 3-bladed H-rotor turbine with a radius of 3.24 m and a blade length of 5 m; Figure 1. The blades are pitched outwards 2 degand have the NACA0021profile with a chord length of 0.25 m at the middle of the blade. The turbine with assembled force sensors is shown in Figure 2 and Figure 3. The sensors are single-axis load cells, which measure tension and compression at a point load. The rotational speed of this turbine can be kept at a constant level [17,19], and the normal force is estimated using the notations from Figure 2 and Figure 3 as the following:

$$F_N = F_0 + F_1 + F_2 + F_3 - F_C \tag{1}$$

where F_C is the centrifugal force:

$$F_C = m\Omega^2 L_C \qquad\qquad (2)$$

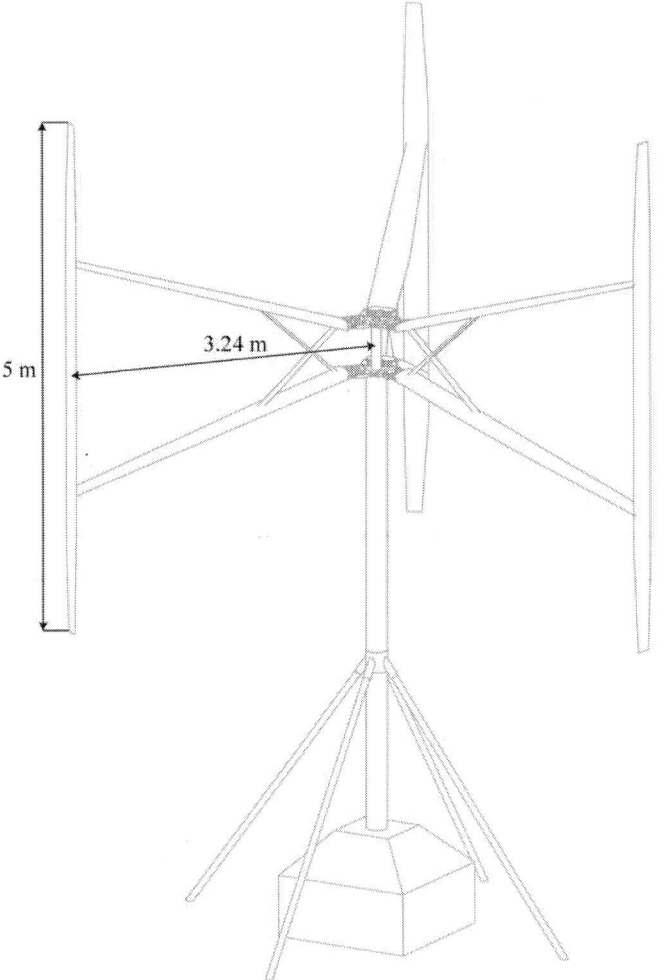

Figure 1. The vertical axis wind turbine (VAWT) used for the experiment.

Here, m=35.79kg is the mass of the blade and support arms, Ω is the turbine rotational speed and L_C=1.83m is the distance from the axis of rotation to the center of mass of the blade assembled with the support arms. F_0, F_1, F_2 and F_3 are the measured forces.

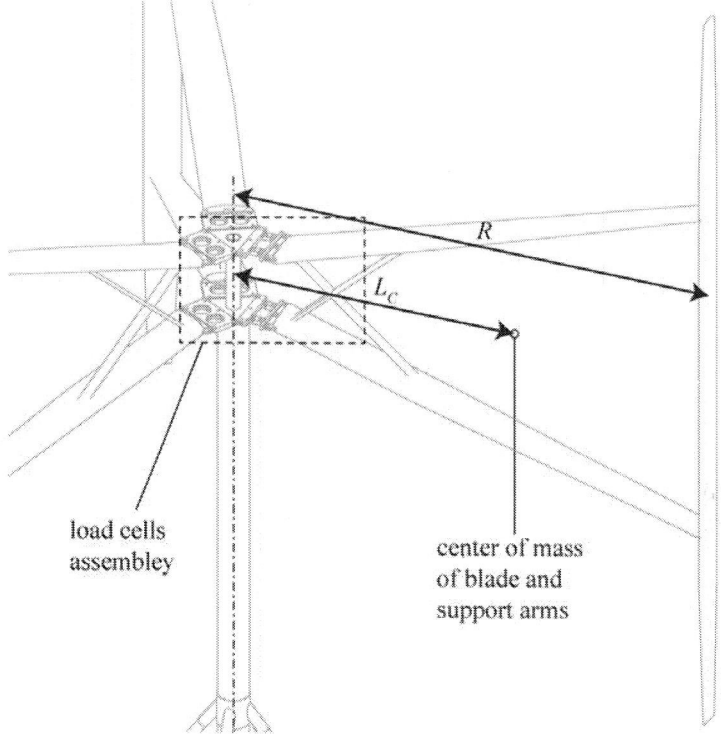

Figure 2. Load cells installed on the VAWT.

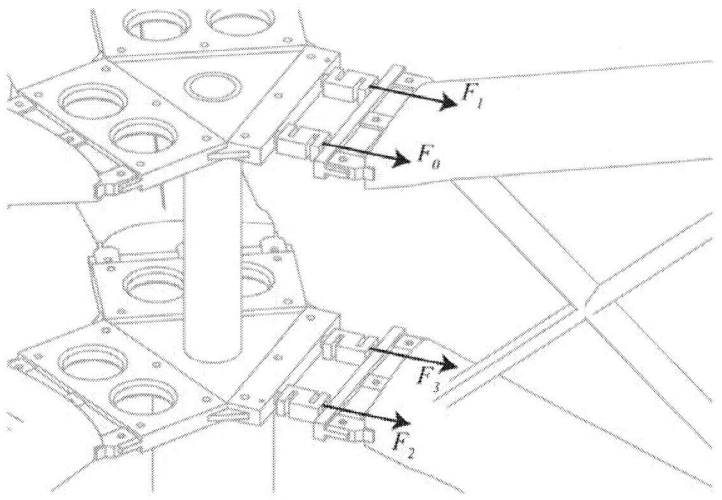

Figure 3. The assembly of the load cells. The notation of the measured forces.

Due to varying weather conditions, the force measurements were analyzed only for times with steady wind flow conditions. The relative standard deviation (RSD) was used to quantify wind flow variations:

$$RSD\left(x\right) = \left(\frac{1}{n}\sum_{j=1}^{n}(x_j - \langle x \rangle)^2\right)^{\frac{1}{2}}\frac{1}{\langle x \rangle} \times 100\% \qquad (3)$$

where $\langle x \rangle$ denotes the average value of variable x.

The measured data were divided into 24 s-long bins: 16 s to stabilize the turbine wake ("wake time", corresponding to 10 revolutions at 40 rpm) followed by 8 s of steady flow operation ("disk time", 5 revolutions at 40 rpm). Wind flow was considered as steady for bins with the RSD of the asymptotic wind velocity V_∞ of $RSD_{wake}(V_\infty){\leq}10\%$, $RSD_{disk}(V\infty){\leq}5\%$ and   the   RSD   of   wind direction Vdir of $RSD_{wake,disk}(V\text{dir}){\leq}1\%$. This definition of the steady wind flow conditions is documented in [18]. Variations of the wind speed during the steady conditions are illustrated in Figure 4.

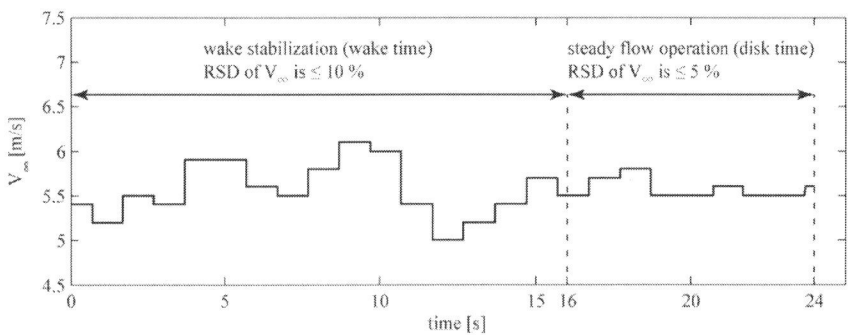

Figure 4. Allowed variations of the asymptotic wind velocity inside a bin with the steady wind flow conditions.

The normal force during one revolution is obtained as the average response over 5 revolutions with steady wind flow. The operational TSR is estimated as:

$$\lambda = \frac{\langle \Omega \rangle R}{\langle V_\infty \rangle} \qquad (4)$$

where Ω is the turbine rotational speed. The average values of Ω and $V\infty$ are taken taken over time with steady wind flow.

The analysis of the measurement accuracy has shown that the maximum error of the measured normal force is a function of the turbine rotational speed:

$$\Delta F_N = \pm \left(0.0049\Omega_{rpm}^2 + 0.072\Omega_{rpm} + 23\right) \tag{5}$$

where Ωrpm is the rotational speed in rpm. For the details regarding the measurement accuracy, the reader is referred to [17,18]. The air density ρ is calculated for the measured air temperature, pressure and humidity according to [17]. The kinematic viscosity v is estimated as the function of the measured air temperature [20].

3. SIMULATION MODEL

This section presents the vortex method together with the dynamic stall model to predict the blade forces of the VAWT. The sign notation of the forces together with the blade azimuth angle are defined in Figure 5.

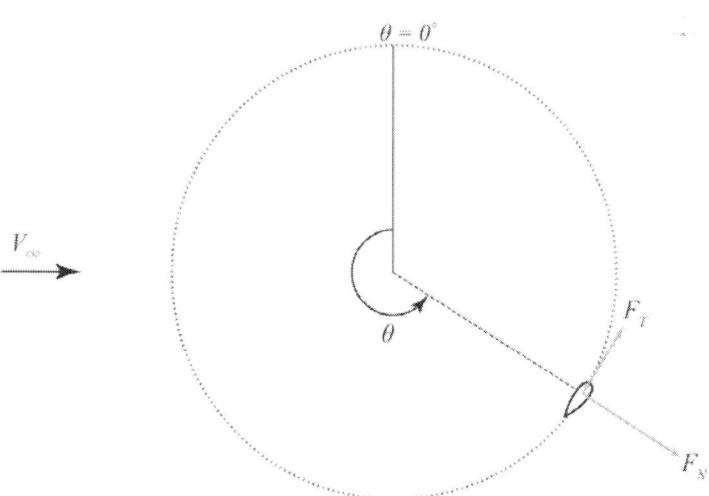

Figure 5. The sign convention of the normal and tangential forces. The counter-clockwise direction of the blade azimuth angle θ is defined as positive.

3.1. Vortex Model of the Turbine

The vortex method is commonly used for solving the flow of vertical axis turbines. The main idea behind the model is to use the vorticity as the discretization variable, instead of the velocity. The vorticity is obtained by taking the curl of the flow velocity:

$$\vec{\omega} = \nabla \times \vec{V} \tag{6}$$

and similarly, the vorticity equation is obtained from the curl of the Navier–Stokes equations:

$$\frac{\partial \vec{\omega}}{\partial t} + \left(\vec{V} \cdot \nabla \right) \vec{\omega} = (\vec{\omega} \cdot \nabla) \vec{V} + \nu \nabla^2 \vec{\omega} \tag{7}$$

which in the two-dimensional case becomes:

$$\frac{\partial \vec{\omega}}{\partial t} + \left(\vec{V} \cdot \nabla \right) \vec{\omega} = \nu \nabla^2 \vec{\omega} \tag{8}$$

The current implementation uses a free vortex model, where the individual vortices are used as discretization variables. The flow velocity is obtained from the model as a superposition of the potential flow solution and the contribution from the vortices:

$$\vec{V} = \nabla \phi + \vec{V}_\omega \tag{9}$$

Under the assumption that the turbine is not confined with walls and that that the blade is approximated as a single point, the potential flow solution $\nabla \phi$ is equal to the asymptotic flow velocity V∞. For the two-dimensional vortex method, according to the Biot–Savart law and using the complex numbers, the velocity contribution from vortices $\vec{V}_\omega$ at position r becomes:

$$V_\omega(r) = \sum_{k=1}^{N_\nu} \frac{i \Gamma_k}{2\pi} \frac{1}{(\vec{r} - \vec{r}_k)} \left(1 - e^{-\frac{|\vec{r} - \vec{r}_k|^2}{\varepsilon^2}} \right) \tag{10}$$

Here, r_k is the position and Γ_k is the circulation of vortex k ($\overline{r}_k$ denotes the complex conjugate of rk); N_v is the total number of released vortices. The Gaussian kernel $\left(1 - e^{-\frac{|r-r_k|^2}{\varepsilon^2}}\right)$ is used to avoid the non-physical divergences when r approaches rk [21].

The relative wind velocity at the turbine blades is a vector sum of the flow velocity at the blade due to its own motion, $-\vec{V}_b$ and the flow velocity $\vec{V}$:

$$\vec{V}_{rel} = \vec{V} - \vec{V}_b \tag{11}$$

where the velocity $\vec{V}$ is given by Equations (9) and (10), and the blade tangential velocity $\vec{V}_b$ is:

$$\vec{V}_b = \Omega R\hat{t} \tag{12}$$

Here, $\hat{t}$ is the unit vector in the tangential direction with the same sign convention as the blade azimuth angle θ in Figure 5. In the Lagrangian formulation, the vortices are allowed to drift with the flow velocity. Neglecting the viscosity outside the boundary layers of the blades, the vortices are propagated according to:

$$\frac{d\vec{r}}{dt} = \vec{V}(\vec{r}) \tag{13}$$

where the velocity $\vec{V}$ is calculated with Equations (9) and (10). The velocities at each vortex position can efficiently be evaluated using the fast multipole method [22], and the current work uses the implementation described in [23].

Even though the vortex method can solve the entire flow, a full solution of the boundary layer will require a high computational effort. To significantly improve the speed, a dynamic stall model of the blade force coefficients is used to obtain the lift and drag coefficients; see Section 3.2. The dynamic stall model requires the flow velocity and the angle of attack to be calculated. The absolute value of the relative flow velocity $V_{rel} = |\vec{V}_{rel}|$ is calculated with Equations (9) to (11) and is used as an input to the blade force model. To properly handle the flow curvature effects from the rotating motion of the turbine, the blade geometry has to be modeled. In the current implementation, the blade is modeled with linear panels with the linear distribution of vorticity according to [5,24]. Using these panels, the bound circulation of the blades Γ_{blade} can be determined by enforcing the no-penetration boundary condition on the surface of airfoil and the Kutta condition on the trailing edge. This circulation can be used to calculate the corresponding angle of attack:

$$\alpha = p \left(\arcsin \left(\frac{\Gamma_{blade}}{\pi c V_{rel}} \right) - \alpha_0 \right) \tag{14}$$

where c is the blade chord length and parameters p and $\alpha0$ are determined by matching the bound circulation Γ_{blade} with the known angle of attack from the steady-state potential flow solutions. The procedure of calculating the angle of attack is described in greater detail in [25], and the method denoted "explicit method" is the one implemented here. This method adds the released vortex into the panel equations and solves for the Kutta condition together with the conservation of circulation, i.e., the total circulation of the released vortex, and the bound circulation should equal the bound circulation from the previous time step.

- With the velocity V_{rel}, the angle of attack α and its time derivative $\alpha\,'$, the lift and the drag force coefficients (C_L and C_D) are obtained within the blade force model for an airfoil profile with a given chord length c and the Reynolds number (Section 3.2, Equation (31)). Please note that only symmetrical four-digit NACA airfoils are implemented in this work. The kinematic viscosity, obtained from the weather data (Section 2), is used in the model to estimate the local Reynolds number. The lift force coefficient from the blade force model can then be used to calculate the effective bound circulation Γ_{ds} with the Kutta–Joukowski lift formula:

$$\Gamma_{ds} = \frac{1}{2} C_L c V_{rel} \tag{15}$$

Due to conservation of circulation, a vortex has to be released from the trailing edge of the blade each time step Δt with a strength $\Gamma_{released}$ corresponding to the change of circulation between the time steps:

$$\Gamma_{released} = \Gamma_{ds,n-1} - \Gamma_{ds,n} \tag{16}$$

where subscripts n and n−1 stand for current and previous time steps. The position of the released vortex is chosen as $0.5\Omega R\Delta t$ behind the trailing edge. In the case of dynamic stall, the bound circulation of the blade will be reduced, i.e., $\Gamma_{ds} < \Gamma_{blade}$. This means that the blade will no longer fulfil the Kutta condition, which then would give infinite flow velocities at the trailing edge. Therefore, during the evaluation of the vortex velocities, it is chosen to approximate the blade as a single point vortex (located at a quarter-chord position) that can be evaluated with Equation (10). This approximation also increases the computational speed (compared to modeling the blades with panels) as no panel to vortex interactions have to be calculated. This approximation is evaluated in [25], and the results show that the difference between approximating the blade with a point vortex or with panels is very small; hence, the approximation is reasonable.

All simulations were performed for 100 turbine revolutions to ensure convergence in the results. This value was chosen from the convergence studies performed in [26]. One hundred twenty time steps were performed for each revolution, and the turbine was kept at a constant rotational speed for the entire simulation. The normal force was calculated based on the obtained lift and drag coefficients:

$$F_N = \frac{1}{2}\rho A_{blade} V_{rel}^2 \left(C_L \cos\varphi + C_D \sin\varphi\right) \qquad (17)$$

with blade area A_{blade} and air density ρ. The angle φ is the angle of the relative flow velocity vector $\vec{V}_{rel}$. The force values are presented from the last revolution of the simulations. An overview of the vortex model algorithm is illustrated in Figure 6, where the most important steps are highlighted.

3.2. Dynamic Stall Modeling

The model originally developed by Leishman and Beddoes [27,28] with modifications for the conditions of VAWTs [29,30] is used for the modeling of the dynamic stall. The dynamic stall model uses experimental data of the lift and the drag coefficients for steady flow over airfoils [31]. This model was combined with the vortex model for a single pitching wing and showed reasonable agreement with experimental data on the pitching airfoils [25]. The inputs to the models are the angle of attack α and its rate of change α', the velocity Vrel, the chord length c, the Reynolds number and the blade airfoil. These inputs are obtained within the vortex model. The outputs of the dynamic stall model are

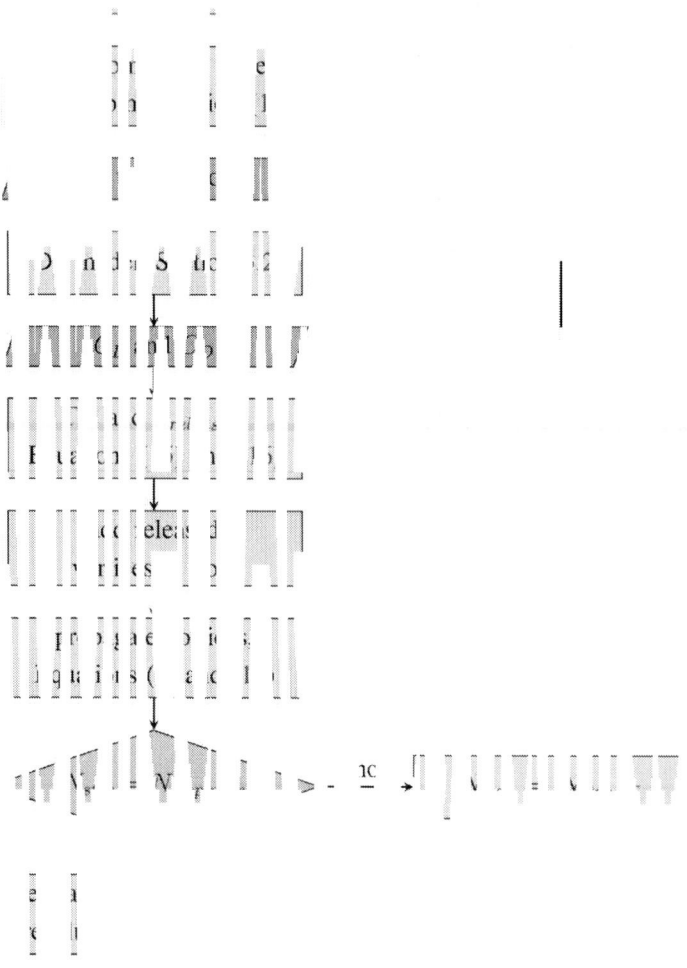

Figure 6. Flow chart of the vortex model combined with the dynamic stall (DS) model. N_{step} is the current time step, and $N_{step,max}$ is the maximum number of time steps. $Nstep,max$=12,000, corresponding to 120 time steps per each of 100 revolutions.

the lift and the drag coefficients (C_L and C_D, respectively). The main principles of the model are described in this section, and the reader is referred to [25,29] for the details, including the empirical parameters.

The Leishman–Beddoes dynamic stall model consists of three parts: unsteady attached flow, dynamic stall onset and unsteady separated flow part. The unsteady attached flow solution comprises impulsive and circulatory loading, which are caused by unsteady boundary vortex and the changes in the angles of attack. The change of the flow velocity due to the vortex contribution is already taken into account by the vortex model. Thus, the unsteady attached flow part of the Leishman–Beddoes model is not used when combined with the vortex model. This method is identical to the one described in [25].

A delay in the pressure response is represented by the further lag in the angle of attack:

$$\alpha'_n = \alpha_n - D_{\alpha_n} \tag{18}$$

where D_α is the deficiency function:

$$D_{\alpha_n} = D_{\alpha_{n-1}} \exp\left(-\frac{\Delta s}{T_\alpha}\right) + (\alpha_n - \alpha_{n-1}) \exp\left(-\frac{\Delta s}{2T_\alpha}\right) \tag{19}$$

Here, T_α is an empirically-derived constant, and its values for the symmetrical NACA-airfoils are found in [29]. For the NACA0021 airfoil, T_α=6.30. The non-dimensional time step Δs is calculated as:

$$\Delta s = \frac{2V_{rel}\Delta t}{c} \tag{20}$$

where V_{rel} is the relative flow velocity obtained from Equation (11).

A critical angle of attack is defined to represent the onset of the dynamic stall:

$$\alpha_{cr_n} = \begin{cases} \alpha_{ds0} & |q_n| \geq q_0 \\ \alpha_{ss} + (\alpha_{ds0} - \alpha_{ss})\frac{|q_n|}{q_0} & |q_n| < q_0 \end{cases} \tag{21}$$

with the reduced pitch rate q as:

$$q_n = \frac{\dot{\alpha}_n c}{2V_{rel}}$$ (22)

Here, q_0 is the reduced pitch rate, which delimits the quasi-steady stall and the dynamic stall; $q_0=0.01$. α_{ds0} and α_{ss} are the critical static stall onset angle and the static stall angle, respectively, $\alpha_{ds0}=17.91°$ and $\alpha_{ss}=14.33°$ for NACA0021 [29]. The dynamic stall condition is defined as when the delayed angle of attack α' exceeds the critical angle of attack α_{cr}:

$$|\alpha'| > \alpha_{cr} \rightarrow \text{stall}$$ (23)

The unsteady separated flow part includes the trailing edge and the leading edge vortex separation. The trailing edge separation is associated with the delay in the convection of the flow separation point over the surface of the airfoil. It is represented via Kirchhoff's flow approximation:

$$f'_n = \begin{cases} 1 - 0.4\exp\left(\frac{|\alpha'_n|-\alpha_1}{S_1}\right) & |\alpha'_n| < \alpha_1 \\ 0.02 + 0.58\exp\left(\frac{\alpha_1-|\alpha'_i|}{S_2}\right) & |\alpha'_n| \geq \alpha_1 \end{cases}$$ (24)

where f' is the delayed separation point and S_1, S_2 and α_1 are the empirically-derived constants, which are based on the local Reynolds number and the airfoil profile and are found in [29]. In addition to the pressure response delay (which is represented by α', Equation (18)), a further delay in the flow separation point is present in order to account for the time-dependent boundary layer:

$$f''_i = f'_i - D_{f_i}$$ (25)

where D_{fn} is:

$$D_{f_n} = D_{f_{n-1}}\exp\left(-\frac{\Delta s}{T_f}\right) + (f'_n - f'_{n-1})\exp\left(-\frac{\Delta s}{2T_f}\right)$$ (26)

Here, $T_f=3$, which is the empirically-derived constant [29]. The normal force coefficient for the unsteady separated flow conditions before the dynamic stall onset is:

$$C_{N_n}^f = C_{N_\alpha} \alpha_n \left(\frac{1 + \sqrt{f_n''}}{2} \right)^2 \tag{27}$$

where $C_{N\alpha}$ is the slope of the normal force coefficient at the static conditions, and it is based on the airfoil and the Reynolds number [29].

When the dynamic stall condition is met (Equation (23)), the leading edge vortex forms and propagates towards the trailing edge and then releases. This vortex convection is represented by an increase in the lift force (sometimes referred to as the vortex lift) during the vortex propagation and followed by a drop in the lift force when the vortex releases. The vortex lift is calculated as the follows:

$$C_{N_n}^v = B_1 \left(f_n'' - f_n \right) V_x \tag{28}$$

where f is the static separation point and B_1 and V_x are parameters, which are based on the local Reynolds number and the blade airfoil and are found in [29].
The total normal force coefficient is the sum of the unsteady normal force coefficient and the vortex lift:

$$C_{N_n} = C_{N_n}^f + C_{N_n}^v \tag{29}$$

An example of the force response during the pitching motion of an airfoil is shown in Figure 7, and the features mentioned above are noted. The tangential force coefficient CT needs to be estimated in order to find the lift and drag coefficients. The calculation of C_T is based on Kirchhoff's approximation, and the dynamic separation point f'' is used:

$$C_{T_n} = \eta C_{N_\alpha} \alpha_n^2 \left(\sqrt{f_n''} - E_0 \right) \tag{30}$$

where η and E_0 are empirical constants, η=0.975 and E_0=0.15 for the NACA0021 profile. When the normal and the tangential force coefficients are known, the lift C_L and the drag C_D coefficients are estimated as:

$$C_{L_n} = C_{N_n} \cos \varphi_n + C_{T_n} \sin \varphi_n \qquad (31)$$

$$C_{D_n} = C_{N_n} \sin \varphi_n - C_{T_n} \cos \varphi_n + C_{D_0} \qquad (32)$$

where C_{D_0} is the drag coefficient at the zero angle of attack and the relative wind flow angle φ is obtained within the vortex model.

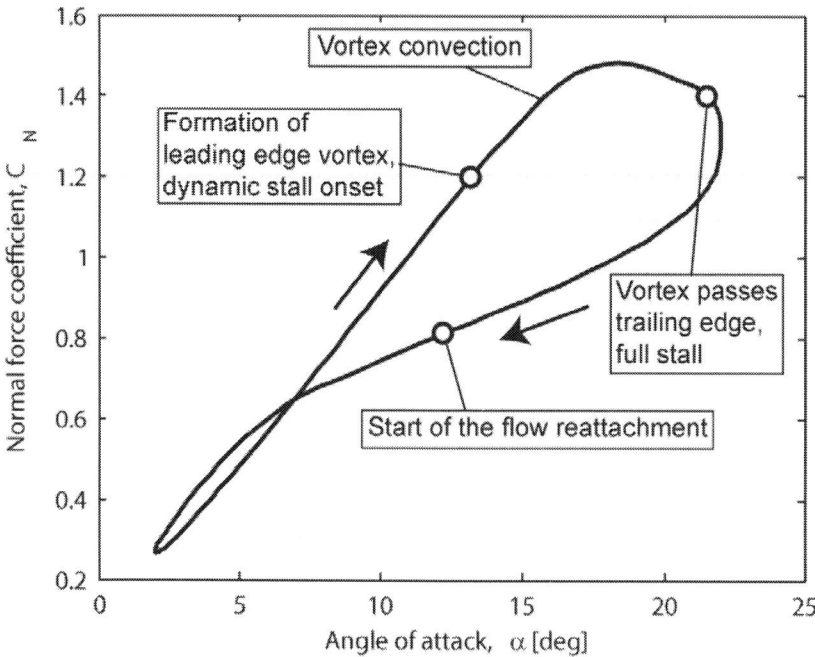

Figure 7. Normal force coefficient for the pitching NACA0021 airfoil, α=12+10sin(ωt), k=0.06, M=0.1, c=0.55m, where k, M and c are the reduced frequency, the Mach number and the chord length correspondingly.

Modification Due to Vortex Shedding

The presented dynamic stall model was tested in [25] for a pitching airfoil, where the flow direction was constant and the blade position was fixed.

However, the blades perform circulatory motion during the operation of VAWTs, and thus, the model has to account for it. A schematic picture of the vortex shedding during the operation of a straight-bladed Darrieus turbine at low TSR is shown in Figure 8. Both the leading and trailing edge vortices are detached and swept away at Quadrant III. Consequently, the delay in the separation is not present in this region. To account for such flow conditions, the dynamic stall model is further modified as in [16]. The delay in the angle of attack and the vortex lift are set to zero to model fast vortex release:

$$\text{quadrant III} \rightarrow \alpha' = \alpha, \ C_N^v = 0 \tag{33}$$

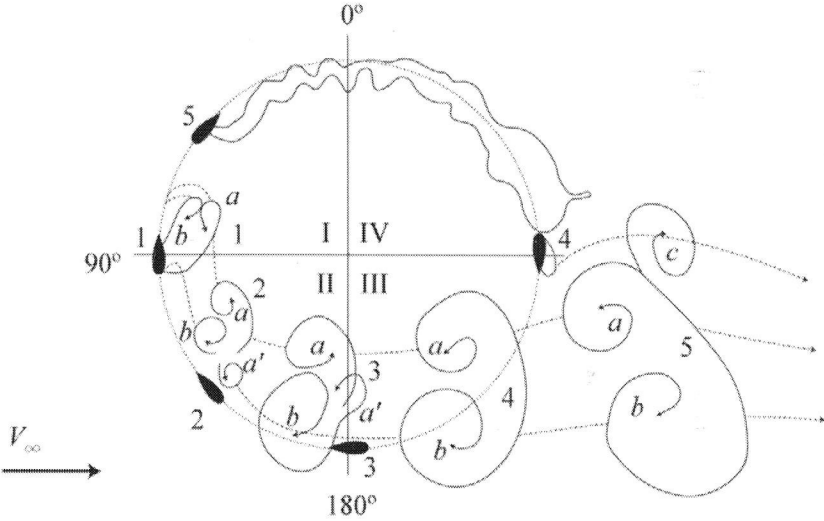

Figure 8. Dynamic vortex shedding for a straight-bladed vertical axis turbine operating in a towing tank at the tip speed ratio (TSR) of 2.14, obtained from [32]. a, a', b and c denote vortices.

4. RESULTS AND DISCUSSION

This section presents the comparison of the simulation results against the measured data at different operational conditions. The discussions regarding the performance of the model are found at the end of this section.

The normal force response at low TSRs is presented in Figure 9, Figure 10 and Figure 11. The maximum magnitude of the F_N-response at the upwind is overestimated at λ=1.84 and λ=2.26, and the shape of the modeled F_N-curve at the downwind deviates from the measurements. The authors presume that the difference between the simulated and the measured values at these low TSRs is due to high magnitudes of the angle of attack. The accuracy of the dynamic stall model decreases with increased angle of attack, which is shown in [25,29], where the dynamic stall model was tested against wind tunnel data for a single blade. As the angle of attack increases with decreased TSR, it is expected that the accuracy of the dynamic stall model should be limited at low TSRs. There is a positive offset of F_N at $\theta=0\circ$, which is mainly due to the blade pitch angle, which was chosen to even out the magnitude of α between the upwind and the downwind regions [18]. The value of the simulated FN-offset is close to the measured one.

Figure 12 shows the F_N-response at $\lambda\approx3$ for two different rotational speeds. The maximum magnitude of the F_N-curve is overestimated for $\Omega=65$rpm, similarly to the overestimation in Figure 9 and Figure 10. The F_N-response at $\lambda\approx3.45$for $\Omega=50$rpm and $\Omega=65$rpm is presented in Figure 13. The results at these conditions are very similar to the results at $\lambda\approx3$, although the model agrees better with experimental data at $\lambda\approx3.45$. The measured F_N-response at $\lambda=3.44$ at $\Omega=65$rpm has a drop in the downwind region at $225\circ<\theta<325\circ$, which is not predicted by the model. The discussions regarding the F_N-drop are found further in this section.

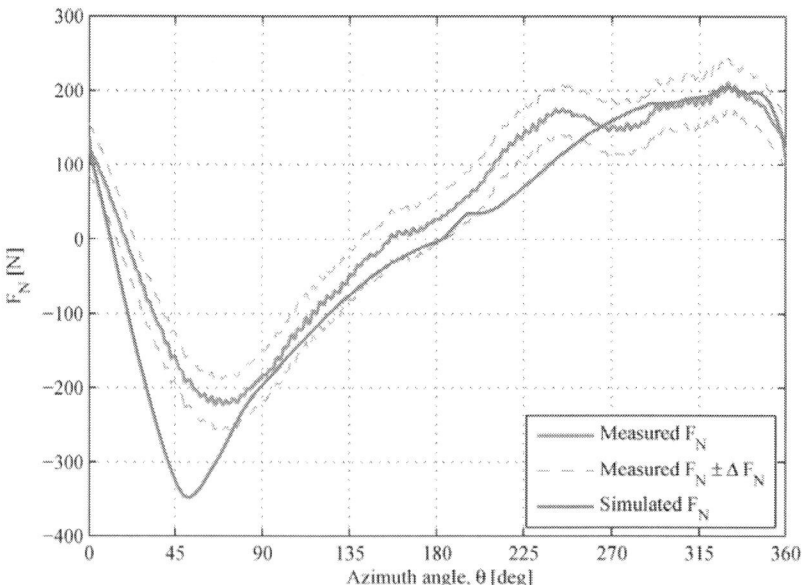

Figure 9. The normal force at $\lambda=1.84$, $\Omega=40.29$rpm. The air density and the kinematic viscosity are $\rho=1.25$kg/m3 and $\nu=1.42\cdot10-5$m2/s.

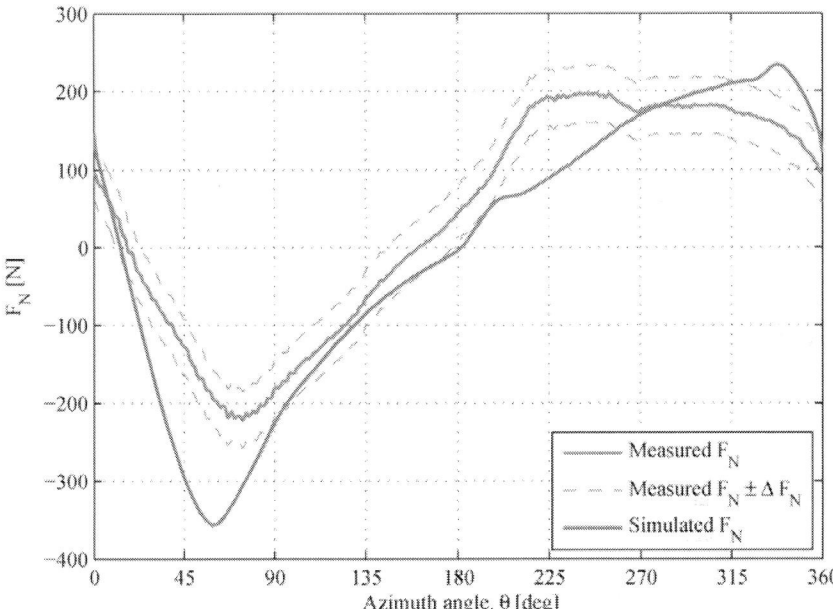

Figure 10. The normal force at λ=2.26, Ω=45.19rpm. The air density and the kinematic viscosity are ρ=1.25kg/m3 and ν=1.42·10−5m2/s.

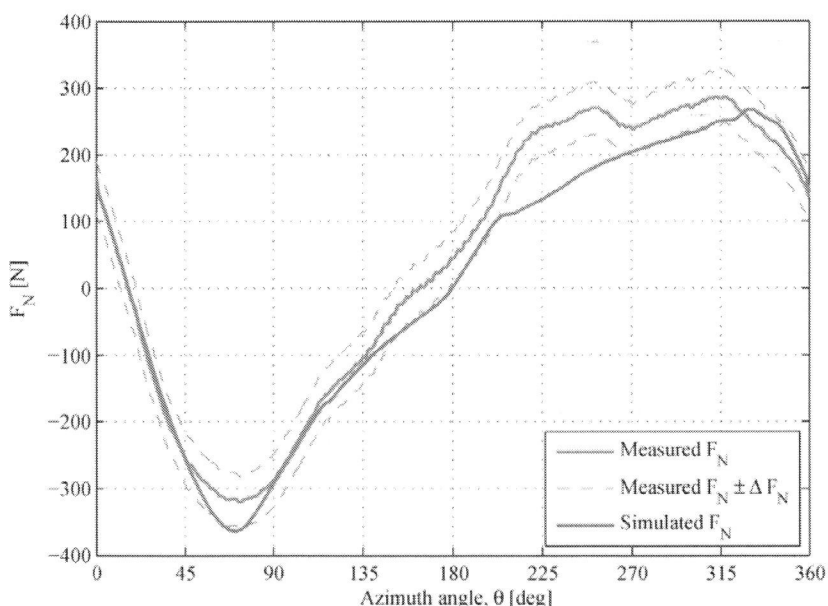

Figure 11. The normal force at λ=2.85, Ω=50.80rpm. The air density and the kinematic viscosity are ρ=1.27kg/m3 and ν=1.39·10−5m2/s.

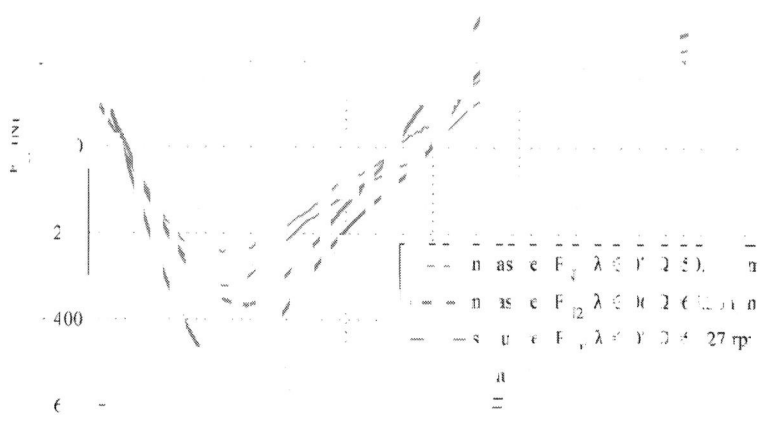

Figure 12. The normal forces at similar λ and different Ω. The air densities and the kinematic viscosities are ρ1=1.27kg/m3, ρ2=1.24kg/ m3 and ν1=1.39· 10−5m2/s, ν2=1.45·10−5m2/s.

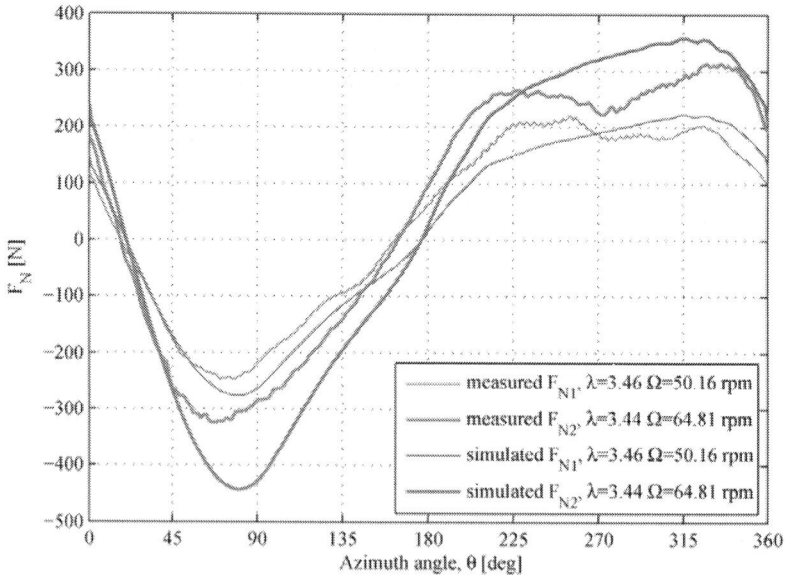

Figure 13. The normal forces at the similar λ and different Ω. The air densities and the kinematic viscosities are ρ1=1.28kg/m3, ρ2= 1.24kg/ m3 and ν1= 1.39·10−5m2/s, ν2=1.45·10−5m2/s.

As the TSR increases, the maximum magnitude of the angle of attack decreases and the prediction of the blade forces becomes more accurate. The F_N-response at $\lambda=3.74$ is shown in Figure 14. The simulated data are in a good agreement with the measured data except the F_N-drop at $235° < \theta < 330°$, which is missed by the model. The FN-response at $\lambda=3.88$ for two different rotational speeds is shown in Figure 15. For both F_N-curves, the F_N-drop in the downwind is not predicted.

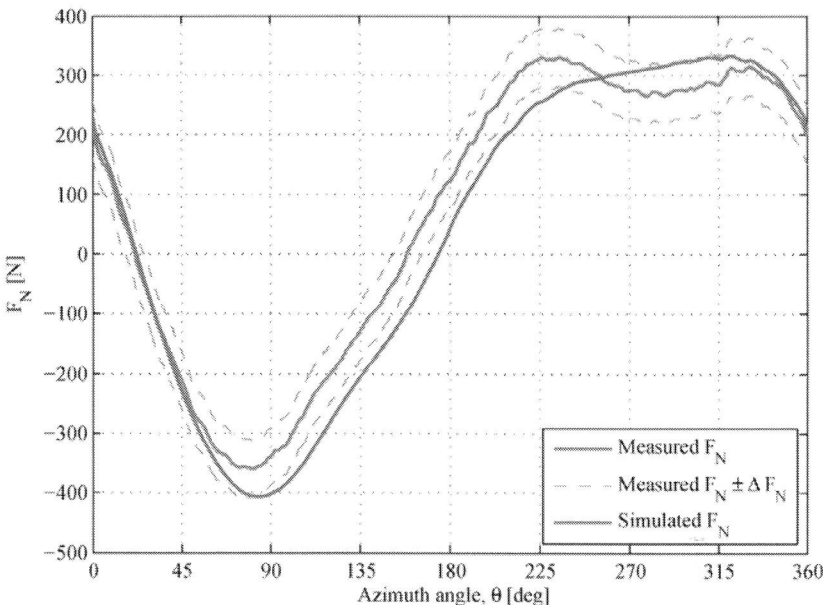

Figure 14. The normal force at $\lambda=3.74$, $\Omega=65.07$rpm. The air density and the kinematic viscosity are $\rho=1.24$kg/m3 and $\nu=1.44\cdot10-5$m2/s.

Two sets of the experimental data with almost identical operational conditions are compared against the simulated results in Figure 16. The shape of the measured F_N-curves is matching, but the magnitudes are slightly different. The maximum difference in the measured F_N magnitudes is $\sim$50N, though the TSR is almost identical ($\lambda1=3.94$ and $\lambda2=4.00$), and the difference in the air density is minor (see the notation to Figure 16). The model shows a close agreement, except that the F_N-drop at the downwind is not present.

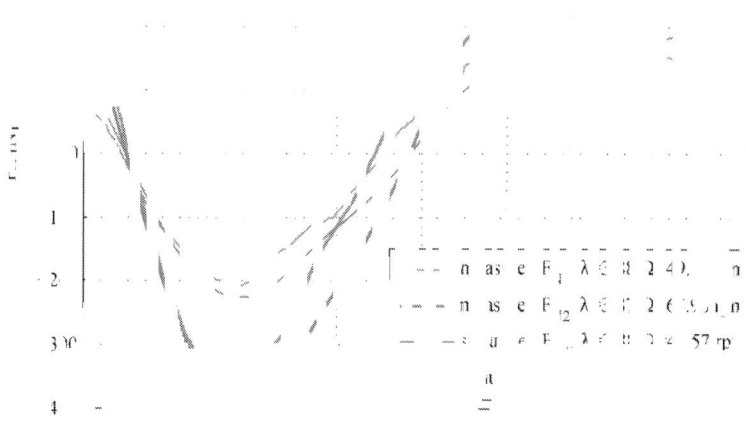

Figure 15. The normal forces at the similar λ and different Ω. The air densities and the kinematic viscosities are ρ1=1.28kg/m3, ρ2=1.24kg/ m3 and ν1=1.39·10−5m2/s, ν2=1.43·10−5m2/s.

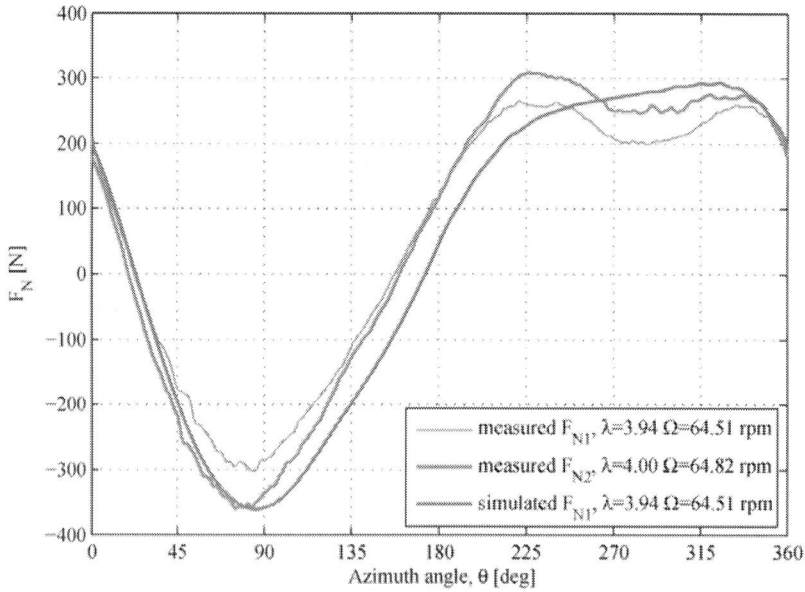

Figure 16. The normal forces for two different sets of data with almost identical λ and Ω. The air densities and the kinematic viscosities are ρ1=1.24kg/m3, ρ2= 1.24kg/m3 and ν1=1.44·10−5m2/s, ν2=1.45·10−5m2/s.

The results at the TSR of 4.6 are presented in Figure 17. At this high TSR, the flow expansion strongly affects the turbine aerodynamics. The model underestimates the maximum magnitude of the F_N-response in the downwind region. The aforementioned F_N-drop at the downwind is clearly observed at λ=4.6, and the model misses it.

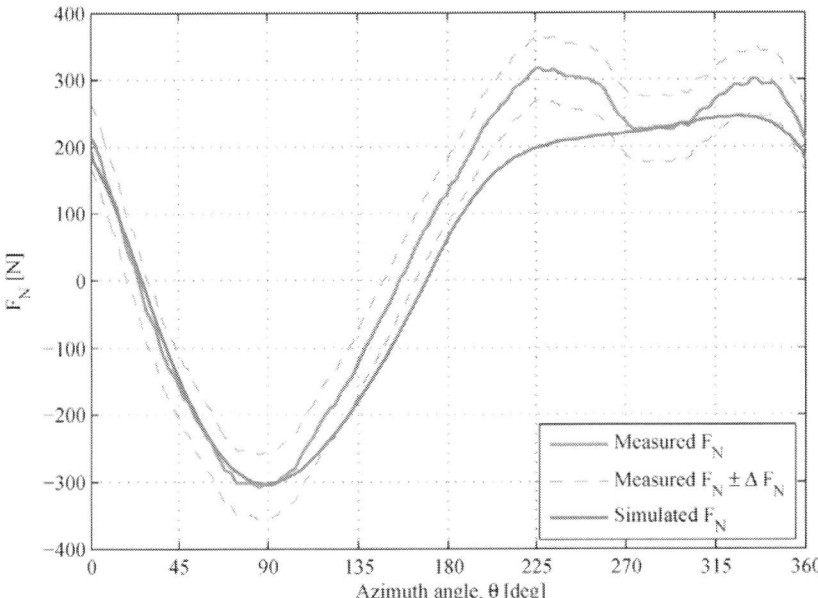

Figure 17. The normal force at λ=4.57, Ω=65.35rpm. The air density and the kinematic viscosity are ρ=1.25kg/m3 and v=1.43·10−5m2/s.

General Discussion

The presented simulations are in 2D, while the measured data are in 3D, and the contribution of the support arms is included in the measured forces. Therefore, it is expected that the presented model cannot reproduce the experimental results in great detail, especially where 3D effects are strong. The flow expansion in the simulation model is limited to the horizontal plane only, and the vertical expansion is omitted. This error should be most prominent at high TSRs, where the flow expansion is largest. Additionally, the current 2D model will not capture wind shear, which would cause a variation of the flow velocity, and hence, the TSR over the turbine height.

Over the whole range of the presented data, the model performs better in the upwind side. This is expected, since the dynamic stall vortex is not implemented in the flow field and the wake effects should be smaller in the upwind side. Furthermore, since the support arms are not included in the model, collision of the blade with vortices from the support arms cannot be reproduced. This is a possible contributing factor to the F_N-drop at $\lambda > 3.4$, as the support arms can have a notable contribution to the wake. The F_N-drop is not expected to be due to the tower wake, since the tower diameter is considerably smaller than the region of the F_N-drop. This drop can also be caused by other three-dimensional effects, such as tip vortices, which are not included in the current model.

There are limitations of the dynamic stall model itself: it is assumed that the blade is a flat plate, and the flow velocity is constant during the change of the angle of attack [29]. Additionally, flow curvature is represented only through a correction in the angle of attack, while it can also influence the empirical constants of the dynamic stall model. These limitations should be considered when evaluating the performance of the model.

The maximum measurement error is estimated for every F_N-curve using Equation (5). Due to the high repeatability of the measured normal force, the shape of the F_N-curve is likely to remain, though the measurement error can change the scale of the F_N-response. This is observed when comparing two sets of data at almost identical operational conditions; Figure 16. Therefore, the measurement error has to be considered throughout the assessment of the simulation model.

The major advantage of the presented model is its computational speed: one simulation with 100 revolutions is in the order of minutes on a single core machine, which is much faster than simulations with 2D CFD models. A 3D vortex model does not have the previously-mentioned constraints with the flow expansion modeling and with the implementation of the support arms. However, the computational time of the existing 3D vortex models is still high, and the computational time of 3D CFD models can be a few months [1]. In this light, the presented simulation model can be used for the fast dimensioning of the turbine loads.

5. CONCLUSIONS

A two-dimensional vortex model for VAWTs was described. The simulation results on the normal forces were assessed against the new experimental data from the straight-bladed VAWT operated at an open site. The comparison is presented for a wide range of operational conditions. There is a drop in the normal force in the downwind region, which is more prominent at high TSR. The authors presume that this drop is due to three-dimensional effects, which are not implemented in the current model. The simulation model shows higher accuracy for the upwind region than for the downwind. At low TSRs, the model misses the measured results, which is expected to be due to the limitations of the dynamic stall model at the high magnitudes of the angle of attack. However, the simulated results agree well with the measured data at moderate and high TSRs, except the force drop in the downwind region. Although the model does not reproduce the experimental results in great detail, it shows a reasonable agreement with experimental data, and it can be used to simulate the maximum load limits on VAWTs at a low computational cost.

ACKNOWLEDGMENTS

This project is conducted with the support of STandUP for Energy. The authors would like to acknowledge the J. Gust Richert foundation for the financial contribution to the equipment for the experiment. Morgan Rossander is acknowledged for experimental work and the estimation of measurement error.

AUTHOR CONTRIBUTIONS

Eduard Dyachuk developed the dynamic stall model in its current implementation, obtained the results from experimental data and wrote the article. Anders Goude developed and described the vortex model and contributed to the data analysis.

REFERENCES

1. Li, C.; Zhu, S.; Xu, Y.L.; Xiao, Y. 2.5 D large eddy simulation of vertical axis wind turbine in consideration of high angle of attack flow. *Renew. Energy* **2013**, *51*, 317–330.

2. Paquette, J.; Barone, M. Innovative offshore vertical-axis wind turbine rotor project. Available online: http://wiki-cleantech.com/wind-energy/innovative-offshore-vertical-axis-wind-turbine-rotor-project (accessed on 1 October 2015).

3. Sutherland, H.J.; Berg, D.E.; Ashwill, T.D. *A Retrospective of VAWT Technology*; Technical Report SAND2012-0304; Sandia National Laboratories: Albuquerque, NM, USA, 2012.

4. Blusseau, P.; Patel, M.H. Gyroscopic effects on a large vertical axis wind turbine mounted on a floating structure.*Renew. Energy* **2012**, *46*, 31–42.

5. Goude, A. Fluid Mechanics of Vertical Axis Turbines. Simulations and Model Development. Ph.D. Thesis, Department of Engineering Sciences, Electricity, Uppsala University, Uppsala, Sweden, 14 December 2012.

6. Islam, M.; Ting, D.K.; Fartaj, A. Aerodynamic models for Darrieus-type straight-bladed vertical axis wind turbines.*Renew. Sustain. Energy Rev.* **2008**, *12*, 1087–1109.

7. Paraschivoiu, I. *Wind Turbine Design—With Emphasis on Darrieus Concept*; Presses Internationales Polytechnique: Montreal, QC, Canada, 2002.

8. Johnston, S.F. *Proceedings of the Vertical Axis Wind Turbine (VAWT) Design Technology Seminar for Industry*; Technical Report SAND80-0984; Sandia National Laboratories: Albuquerque, NM, USA, 1982.

9. Akins, R.E. *Measurements of Surface Pressures on an Operating Vertical-Axis Wind Turbine*; Technical Report SAND89-7051; Sandia National Laboratories: Albuquerque, NM, USA, 1989.

10. Oler, J.; Strickland, J.; Im, B.; Graham, G. *Dynamic Stall Regulation of the Darrieus Turbine*; Sandia National Laboratories: Albuquerque, NM, USA, 1983.

11. Strickland, J.H.; Webster, B.T.; Nguyen, T. A vortex model of the Darrieus turbine: An analytical and experimental study. *J. Fluids Eng.* **1979**, *101*, 500–505.

12. Ashuri, T.; van Bussel, G.; Mieras, S. Development and validation of a computational model for design analysis of a novel marine turbine. *Wind Energy* **2013**, *16*, 77–90.

13. Shires, A. Development and Evaluation of an Aerodynamic Model for a Novel Vertical Axis Wind Turbine Concept.*Energies* **2013**, *6*, 2501–2520.

14. Keinan, M. A Modified Streamtube Model for Vertical Axis Wind Turbines. *Wind Eng.* **2012**, *36*, 145–180.

15. Wang, K.; Hansen, M.O.L.; Moan, T. Model improvements for evaluating the effect of tower tilting on the aerodynamics of a vertical axis wind turbine. *Wind Energy* **2015**, *18*, 91–110.

16. Dyachuk, E.; Goude, A. Simulating Dynamic Stall Effects for Vertical Axis Wind Turbines Applying a Double Multiple Streamtube Model. *Energies* **2015**, *8*, 1353–1372.

17. Rossander, M.; Dyachuk, E.; Apelfröjd, S.; Trolin, K.; Goude, A.; Bernhoff, H.; Eriksson, S. Evaluation of a blade force measurement system for a vertical axis wind turbine using load cells. *Energies* **2015**, *8*, 5973–5996.

18. Dyachuk, E.; Rossander, M.; Goude, A.; Bernhoff, H. Measurements of the Aerodynamic Normal Forces on a 12-kW Straight-Bladed Vertical Axis Wind Turbine. *Energies* **2015**, *8*, 8482–8496.

19. Kjellin, J.; Bülow, F.; Eriksson, S.; Deglaire, P.; Leijon, M.; Bernhoff, H. Power coefficient measurement on a 12 kW straight bladed vertical axis wind turbine. *Renew. Energy* **2011**, *36*, 3050–3053.

20. Co, C. *Flow of Fluids Through Valves, Fittings, and Pipe*; Number 410; Crane Company: Joliet, IL, USA, 1988.

21. Beale, J.T.; Majda, A. High order accurate vortex methods with explicit velocity kernels. *J. Comput. Phys.* **1985**, *58*, 188–208.

22. Greengard, L.; Rokhlin, V. A Fast Algorithm for Particle Simulations. *J. Comput. Phys.* **1987**, *73*, 325–348.

23. Goude, A.; Engblom, S. Adaptive fast multipole methods on the GPU. *J. Supercomput.* **2013**, *63*, 897–918.

24. Ramachandran, P.; Rajan, S.C.; Ramakrishna, M. A fast, two-dimensional panel method. *SIAM J. Sci. Comput.* **2003**, *24*, 1864–1878.

25. Dyachuk, E.; Goude, A.; Berhnoff, H. Simulating pitching blade with free vortex model coupled with dynamic stall model for conditions of straight bladed vertical axis turbines. *J. Sol. Energy Eng.* **2015**, *137*, 041008.

26. Goude, A.; Ågren, O. Simulations of a vertical axis turbine in a channel. *Renew. Energy* **2014**, *63*, 477–485.

27. Leishman, J.G.; Beddoes, T.S. A generalised model for airfoil unsteady behaviour and dynamic stall using the indicial method. In Proceedings of the 42nd Annual Forum of the American Helicopter Society, Washington, DC, USA, 2–5 June 1986; Westland Helicopters Ltd.: Yeovil, UK, 1986; pp. 243–265.

28. Leishman, J.G.; Beddoes, T.S. A semi-empirical model for dynamic stall. *J. Am. Helicopter Soc.* **1989**, *34*, 3–17.

29. Dyachuk, E.; Goude, A.; Bernhoff, H. Dynamic Stall Modeling for the Conditions of Vertical Axis Wind Turbines.*AIAA J.* **2014**, *52*, 72–81.

30. Sheng, W.; Galbraith, R.A.M.; Coton, F.N. A Modified Dynamic Stall Model for Low Mach Numbers. *J. Sol. Energy Eng.* **2008**, *130*, 1–10.

31. Sheldahl, R.E.; Klimas, P.C. *Aerodynamic Characteristics of Seven Symmetrical Airfoil Sections through 180-Degree Angle of Attack for Use in Aerodynamic Analysis of Vertical Axis Wind Turbines*; Technical Report

SAND80-2114; Sandia National Laboratories: Albuquerque, NM, USA, 1981.

32. Brochier, G.; Fraunié, P.; Béguier, C.; Paraschivoiu, I. Water Channel Experiments of Dynamic Stall on Darrieus Wind Turbine Blades. *AIAA J. Propul. Power* **1986**, *2*, 445–449.

CHAPTER 9

The Flight of Birds and Other Animals

Colin J. Pennycuick

School of Biological Sciences, University of Bristol, Bristol BS8 1TQ, UK

ABSTRACT

Methods of observing birds in flight now include training them to fly under known conditions in wind tunnels, and fitting free-flying birds with data loggers, that are either retrieved or read remotely via satellite links. The performance that comes to light depends on the known limitations of the materials from which they are made, and the conditions in which the birds live. Bird glide polars can be obtained by training birds to glide in a tilting wind tunnel. Translating these curves to power required from the flight muscles in level flight requires drag coefficients to be measured, which unfortunately does not work with bird bodies, because the flow is always fully detached. The drag of bodies in level flight can be determined by observing wingbeat frequency, and shows C_D values around 0.08 in small birds, down to 0.06 in small waders specialised for efficient migration. Lift coefficients are up to 1.6 in gliding, or 1.8 for short, temporary glides. In-flight measurements can be used to calculate power curves for birds in level flight, and this has been applied to migrating geese in detail. These typically achieve lift:drag ratios around 15, including allowances for stops, as against 19 for continuous powered flight. The same calculations, applied to Pacific Black-tailed Godwits which start with fat fractions up to 0.55 at departure, show that such birds not only cross the Pacific to New Zealand, but have enough fuel in hand to reach the South Pole if that were necessary. This performance depends on the "dual fuel" arrangements of these migrants, whereby they use fat as their main fuel, and supplement this by extra fuel from burning the engine (flight muscles), as less power is needed later in the flight. The accuracy of these power curves has never been checked, although provision for stopping the bird, and making these checks at regular intervals during a simulated flight was built into the original design of the Lund wind tunnel. The *Flight* programme, which does these comparisons, also had provision for including contributions due to extracting energy from the atmosphere (soaring), or intermittent bounding flight in small birds (Passerines). It has been known for some time that the feathered surface allows the bird to delay or reverse detachment of the boundary layer, although exactly how this works remains a mystery, which might have practical applications. The bird wing was in use in past times, when birds were still competing with pterosaurs, although these had less efficient wings. The birds that survived the extinction that killed the pterosaurs and dinosaurs have (today) an automatic spherical navigator, which enables them to cross the Pacific and find New Zealand on the other side. Bats have never had such a device, and pterosaurs probably did not either. Animals, when seen from a zoological point of view, are adapted to whatever problems they had to deal with in earlier times.

KEYWORDS

birds; bats; pterosaurs; aerodynamics; migration; wind tunnel

1. THE BIOLOGIST AS REVERSE ENGINEER

Biologists look at birds as creatures that do things that we cannot. In the 21st century we can see what they are doing with their wings, but until the 20th century that was unimaginable. We know that birds can fly, but we are not engineers, and do not try to modify them to fly better. We biologists are, in effect, reverse engineers. We know nothing about the niches that birds fit into or about the constraints that limit what is possible, with biological material. In the light of engineering, we begin to see what can be done, and start to look in detail at how birds do things that we did not originally expect, like leaving Alaska and spending a week flying non-stop to New Zealand, or using natural thermal convection to fly from Europe to Africa and back, for no cost at all. We can use satellite tracking to see what birds do when they migrate, but as to what they are "designed" to do, then we are biologists, and we have to find out what they do first, before asking how they do it. We do not initially know what limits their performance, what pushes them so hard to do better, what their starting points were, or what pushes them so hard to do things that no animal has done before. Animals have been flying for 200 million years, and birds are not the first to succeed, or the only ones to try. However, we know now how to build a wind tunnel, so we built one, taught pigeons to fly in it, and carry on from there.

2. ANIMALS IN THE WIND TUNNEL

2.1. Layout Limitations

Wind tunnel engineers are used to working with models of reduced size, flown in conditions that are not the same as in the real world. Those problems do not arise when testing birds, because if you select a bird like a pigeon, you can fit the whole of it in the test section, and set the conditions as they are out of doors. That is a great convenience, because the pigeon does not have an anatomy in the way than an engineer can understand it. If you hold a pigeon in the hand, its wings are folded up, and if you unfold them, they can be set at dihedral angles from −60 to +90 degrees, and a span between one-third and the whole span, with all kinds of angles that can be set by muscles to a wide range of values. A pigeon wing cannot be set by hand into a position that is meaningful for flight. On the other hand, the whole pigeon can be trained to fly in a wind tunnel, and will set itself into a posture for whatever kind of flight it has to perform. You do not measure forces in a bird wind tunnel. You let the bird set itself up, and infer the forces from the geometry of its situation.

Bird wind tunnels are designed to achieve given levels of performance without being too expensive. Engineers, when consulted about this, often say that the best way is an open-circuit suction tunnel, with the fan at the downstream end. This type of design works very well when used in the wind tunnel laboratory, where the test section (containing air below ambient pressure) is carefully sealed. However, the ornithologist cannot leave his bird unattended. He needs access to it, and consequently leaves holes and doors open, which allows outside air to rush invisibly in, and disturb the flow in the test section. A suction tunnel does not work correctly, if air can get into the test section by routes not noticed in the design. In practice, blower and recirculating tunnels can be used [1,2] but suction tunnels are not practical for any form of quantitative work on live birds.

Figure 1 shows an open-circuit blower tunnel used in Bristol in the 1960s and later moved to Nairobi. The bird flies in the open air, at the exit from the contraction, which is at the current environmental static pressure, and allows the experimenter direct access to the bird or bat, without interfering with the air flow. This tunnel may have had some shortcomings, but it worked very well to get glide polars of pigeons, and allowed the range of lift and drag coefficients in the normal flight of birds to be determined. This tunnel was designed to allow forces and work to be scaled, in a set of similar animals in the manner pioneered by the physiologist A.V. Hill in the 1930s [3], although his approach was not followed by later physiologists after the war.

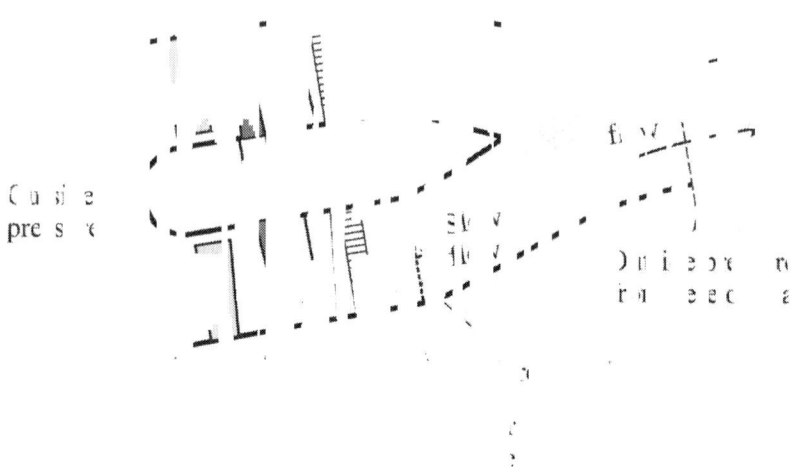

Figure 1. Hydraulic-powered wind tunnel at Bristol in 1966, moved to Nairobi in 1970. This tunnel was powered by a Woods ventilating fan, with a contraction ratio of about 4. Fixed stator vanes before and after the fan straightened the flow, and there was a honeycomb and a fine mesh screen before the contraction. The pigeon flew outside the end of the contraction, where it could be photographed and measured at will. Maximum speed 22.1 m/s.

A better (but more expensive) solution was built by AB Rollab of Solna, Sweden, and installed in the Biology Department of Lund University in 1994 (Figure 2) [4]. This is a closed-circuit wind tunnel with a test section measuring 1.20 m across by 1.08 m vertically, and a contraction ratio of 12.25:1, rigged in such a way that the entire machine, return flow and all, can be tilted from 8° descent to 6° climb. The first 1.20 m of the test section is enclosed by walls, then there is an open 0.5 m gap before the line to the motor begins. The rest of the tunnel is entirely enclosed, so that the lowest pressure is at the test section, and air in the rest of the circuit is at higher pressure. This means that the experimenter can move equipment in and out of the gap in the test section, or reward the bird for its performance, without disturbing the flow where the bird is. Another version of this tunnel was installed by the same Swedish manufacturers later at the Max Planck Institute for Ornithology at Seewiesen in Germany, but this one has a horizontal circulation, and omits some of the original features, notably the tilting facility.

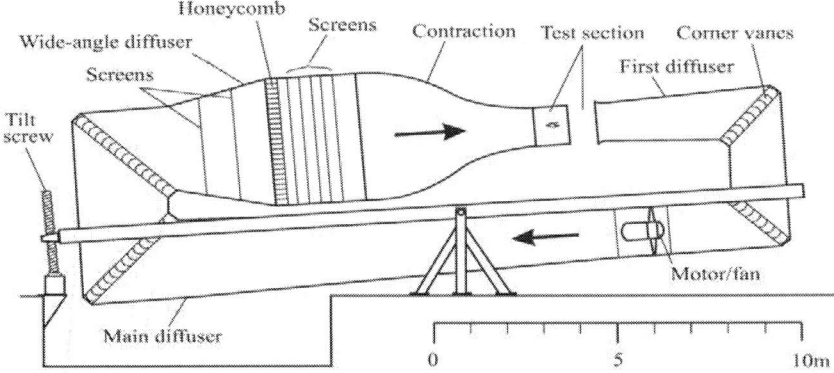

Figure 2. The Lund wind tunnel, installed in 1994, has a test section measuring 1.20 × 1.08 m and a contraction ratio of 12.25. There is a honeycomb, 5 screens and a fine wire screen. The tilting arrangement allows for 8° climb to 6° descent. The closed-circuit arrangement is open at the test section, but closed everywhere else. Maximum speed 50 m/s.

2.2. Gliding Flight

The simplest way to use a bird to make its own measurements is to train it to hold a constant position in a wind tunnel, in which the air flow has been set by the experimenter. For example, just about any bird can be trained to fly in a wind tunnel, in a body of air that is moving slightly upward relative to horizontal, at a speed that is within its normal range [5]. By doing this over a range of speeds, and finding the minimum descent angle at which the bird can fly without flapping its wings, the basic "glide polar" can be found for the bird. This curve can be generalised, and used for gliding calculations in a wide range of birds. Unlike an engineer, you do not have to worry about small amounts of drag caused by poor wing-to-body junctions, because all birds are constantly modified by natural selection to deal with small performance defects, caused by minor variations of anatomy. Once the shape and size of wing are set to suit the bird's life style (also by natural selection) the bird is optimised as well as it can be, to maximise performance. This type of optimisation only optimises species that already exist, and works over a period of many generations, but that is, of course, what biologists have to deal with.

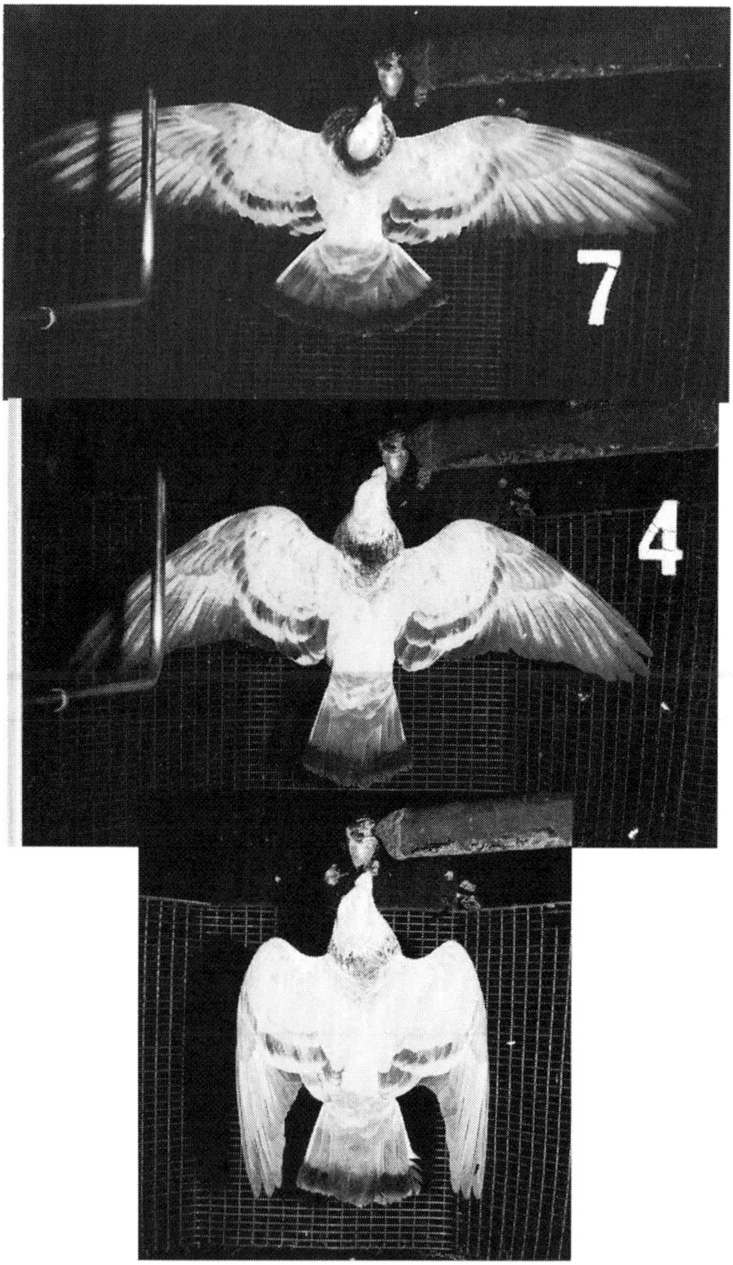

Figure 3. A pigeon gliding in the Bristol wind tunnel, showing planform variations with speed. There are no mechanical constraints with this in a bird wing, but there are in bat or pterosaur wings. The lift coefficient went from 1.3 to 0.3, not taking account of the tail area. It would have been much lower at high speeds, if the full span had been maintained.

A gliding pigeon (Figure 3) is quite a spectacle for a glider designer. In the first place, it gets unsteady at low speeds, but still flies nearly as well as before, if you put a rubber band around its tail to prevent it from spreading. How, then does it control its speed? The pigeon's wings can move about the shoulder joints, forwards or back, upwards or down, and twisting nose-down or nose-up. To control speed, it moves them forward or back, like a hang-glider pilot. Moving the wings forward also spreads them to their greatest area, while moving them back allows the elbow and wrist joints to flex, so that the feathers of the wing slide inwards over one another, reducing the span and area of the wing. Fast-gliding birds can reduce their wing area and span to less than half the maximum values, without losing any strength whatsoever. This is possible because the bird wing is a multiple-spar design, with the shaft of each feather being a local spar for its own part of the wing. The same movement that adjusts the area of the wing also serves to control the area, by allowing the feathers to slide over one another. Varying the span is a better method of adjusting wing planform for speed than adjusting the chord, for instance by Fowler flaps. However, it involves a set of elbow and wrist joints, whose axes are out of line with each other, to allow various parts of the wing to be aligned, in a way that is unfamiliar in aircraft wings.

2.3. Measuring Drag

To generalise a measured glide polar, that is, adapt a measured polar to get new polars for other birds, you can add up sources of drag, starting with the drag of the feathered body, minus wings. If you try that, you get into trouble straightaway. If you take a dead bird, or stuffed or simulated one, no matter what you do, the air flow just separates from the body surface, and will not reattach. It gives a drag coefficient of 0.3 or more, like a football. As with all wind tunnel measurements on birds, trying to measure drag on parts of a dismembered bird is useless.

This problem was not solved until a way was found to measure a living bird's body drag, without balances or any form of external measurement. It was noticed some years ago that this body drag coefficient is one of two morphological variables needed to calculate the speed (V_{mp}) at which a particular bird requires minimum power to fly level. The other one is the dimensionless variable k, which accounts for differences in lift distribution from the ideal ellipse. k can be got at from field observations of wild birds and is a little below 1, because of splayed feathers at the wing tips. The other variables for the calculation are easily measured, like wing span, gravity, air density and so on. The experiment for the body drag coefficient is to fly the bird level, and measure its wing beat frequency, where it is easy to show that the small variations of frequency vary directly with the power. If the drag coefficient is high, V_{mp} comes down, and if it is low V_{mp} goes up. You take a few dozen of those frequency curves, from the minimum to the maximum speeds at which the bird will fly horizontally, and having measured V_{mp}, you turn the equation round to get the body C_D. A bird with no back load has a value around 0.1 or a bit less, but adding even small bits of plastic on the bird's back sends its C_D up sharply [6]. That is a serious problem for those of us who use satellite tracking to follow long-distance migrants, as we do not have an easy way to check increases of body drag, caused by add-on equipment boxes.

2.4. Power in Biology

A bird flying horizontally overcomes drag by flapping its wings about the shoulder joint. The flapping motion results in a mean horizontal thrust force, and this force, multiplied by the speed, is the power required from the flight muscles [6]. The power comes from the stress developed by the muscle, and the speed at which it can shorten, and these two variables set the power available for shortening in the muscles of birds of different size. Measuring a bird's total drag is the starting point in any discussion of power requirements in level flight, but the drag itself is not a simple calculation, as in a fixed wing.

To measure the drag, you have to measure (or estimate) the rates of all processes in the living bird that require energy from fuel reserves. This begins with the basal metabolism, the power needed to keep the bird alive, which scales roughly with the 0.75 power of the body mass in birds of different mass. The mechanical power required to fly, neglecting comparisons, varies with about the 1.17 power of the mass. An African vulture requires less than 4% of its cruising power to stay alive, whereas a small passerine like a Goldcrest uses over 30% of its power for the same purpose. Also the small bird goes slower than the large one. Consequently small birds cannot afford any kind of migration strategy that wastes time, whereas large birds can use soaring procedures that cut their energy expenditure, at the expense of a lot of time spent soaring in thermals. The energy used in flapping the wings, and overcoming the drag of the wings also has to be expressed in terms of the fuel energy consumed, rather than of the work done. The task of calculating the power needed by a flapping wing is far beyond current experimenters, but can be handled by analogy methods, relating this work to the amount of work that would have to be done, without flapping the wings. The details of how that is done are in my book *Modelling the Flying Bird* [2].

Biologists know that using muscles to generate mechanical power results in the consumption of fuel, but they call this "metabolism", and think of it as something that occurs during muscular activity, but not connected with other changes that occur in flight. They never measure the air density, or think about changing gravity. If the bird is "flying", that is enough, and if it does manoeuvres like cyclical acceleration and deceleration, they are just ignored. Consequently, published measurements of the rate of fuel consumption when flying, which have been made by physiological methods, are not very helpful to the flight theorist. The key to using this approach for practical calculations is to define a speed V_{mp}, at which the bird requires less power to fly, than one flying faster or slower, and devise a practical way to measure it, which can be done by measuring wingbeat frequency in level flight.

Biologists like to do statistics on every kind of wing that is of interest to them, while those of us who study flight behave more like engineers, and define any bird by its mass, wing span and wing area, plus those environmental variables that affect the result, *i.e.*, gravity and the air density. Having defined these primary variables, we can define others that follow, and combine the whole thing into a programme called *Flight*, which has been available from [2] for many years, and is essentially a model of a flying bird. If we have a bird and we know its measurements and the air density, then we also have a power curve for it, *i.e.*, a curve that relates power required, and rate of loss of fuel, to the forward speed. The curve is severely constrained by natural selection, as we can see by thinking about birds that do not exist. If we extend our known measurements upwards to birds bigger than any that we know (20 kg and up), we find that the wingbeat frequency goes down, and the work done by each gram of muscle in each contraction goes up, until eventually the bird cannot fly at all, because the requirements for stress and strain are above the upper limit for flight muscles. There are no huge flapping birds. Beyond a certain size, there is no room for extra muscle to fly at V_{mp}, and well before that, the bird cannot lift extra load in the form of fuel, or do sudden manoeuvers to change speed or avoid obstacles. Current birds fly up to a body mass around 16 kg, where they have obvious difficulties, and there are none at all above 20 kg.

There was once a group of very large flapping animals, the last known pterosaurs of the Cretaceous period. At the end, for a short time, they had wing spans approaching 12 m, far more than earlier pterosaurs. We do not know what happened, but most probably the earth changed, by changing minerals in the mantle into a less dense form, and moving them outwards. This would reduce the surface gravity without changing the Earth's mass [7]. If we extend the measurements down to miniature hummingbirds (3 g and down) the wingbeat frequency gets so high that the wing muscles do not have time to get ready for the next contraction, a problem that insects overcome with a different type of muscle, that develops low-amplitude contractions in the kilohertz range. Meanwhile, the regular-sized birds all have essentially the same anatomy, with adjustable mass, wing span and wing area. They obey the rules, and adjust everything together. The best combination of power and wingbeat frequency is found in medium-sized birds, especially waders of around 300 g, some of which fly over 10,000 km non-stop. Range performance fades to nothing in very small and very big birds.

2.5. Accuracy of Calculations

These arguments depend on calculating the power required in level flight, which we can do, but we can only check the accuracy of the results by looking at the performance of migrants. We got a chance to try this in 2008, when we were asked by the BBC to help tracking geese on their spring migrations [8]. We joined in a BBC radio project by Julian Hector, called "World on the Move" in which a great variety of animals were tracked as they migrated, and our job was to work out each goose's remaining fuel, and show it on the website. When the birds were airborne, we had the help of the BBC weather department to get the winds and the groundspeeds. The GPS tracks gave the ground speeds, which were sometimes so low that it was clear that the birds had stopped, although this was not visible from the GPS points. We could calculate the power the geese needed to fly at the measured air speed, or sit on the ground, and the mass of fuel consumed.

Figure 4 shows the routes that the different kinds of geese followed. We saw the climb up the Greenland ice cap, the increased speed at the top, and the fact that the big geese (Whitefronts) had trouble flying when they were up there and had to walk at times, while smaller geese (Brents) could fly over the top, and carry on up into Canada on the west side. We had estimates of starting fuel loads, and saw that these geese always arrive anywhere with at least 200 km of *energy height* in hand (more later), except when going south in autumn, when they crept into the wintering area without much fuel in hand. In other words, the predicted fuel amounts from the *Flight* programme gave quite a detailed idea of exactly how the geese set up their migratory flights, and showed that they did not stop for head winds, but just carried on, head wind or not. However, we had no direct way of checking whether our calculated fuel accounts were right, as we could not get at the birds when they were out in the wilds.

In its predictive format, the *Flight* programme prints out a list of the current values of 30 variables that change as the bird goes along, starting with the mass, the speed and the power. If we were to catch the bird over the ocean, those figures say what we would expect to see, if we were to catch and inspect the geese. Actually, the Lund wind tunnel was originally designed to make those checks possible, during a long flight in the wind tunnel. To do that, you have to swing a balance into the test section every hour or so, and stop the wind for a minute or two so that the bird can weigh itself. You also have to make provision so that the bird can control its flight speed, using sensors that determine whether the bird is creeping to the upstream or downstream end of the test section, and adjusting the speed accordingly. This arrangement of sensors and speed control was included in the original design [4], but it would require a lot of time to get all that working. The tunnel is all set up to monitor a flight lasting days, allowing the bird to set its speed, and keeping track of its weight, measuring fuel, speed and everything else as we go, but this has not yet been done.

3. MIGRATION IN SMALL BIRDS

Small birds have much the same density as bigger ones, and scaling effects see to it that their speeds for minimum power (V_{mp}) and maximum range (V_{mr}) can be calculated as in bigger birds. The winds are the same for any bird, and a goose that can cruise at 20 m/s does a lot better than a sparrow going at 10 m/s, in the same winds. Another general effect of scaling laws is that big birds have higher aspect ratios, like 9 for a goose, whereas little birds that flit about in the bushes typically have aspect ratios of 5 to 7. That means that if you think of migration range in terms of energy height (more below), the distance that each bird goes depends on its remaining energy height, and its lift:drag ratio. That is Breguet's Law, and nothing else is involved. Small birds in general have smaller lift to drag ratios than large ones, and waste more energy on their basal metabolism, a biological energy ratio which is not otherwise needed in the range calculations themselves. Low cruising speed is a major waster of energy for small birds, as they use energy for basal metabolism (staying alive), and the smaller the bird, the bigger the fraction of the total power needed for that. They need to do something that allows them to increase their cruising speed, which will save on basal metabolism, without a big penalty in lift:drag ratio.

Scaling laws also show that swans can hardly find enough muscle power to fly, whereas blackbirds, whose muscles shorten at a higher frequency, have plenty of power for vertical take-off, sudden changes of direction and so on. Does a blackbird cruise along, using only half its flight muscle? Rather than that, it uses the whole muscle, at a higher frequency, but only for part of the time. The characteristic flight style of small birds (passerines) is "bounding", in which the bird flaps for a few wingbeats, then closes its wings completely, and follows a parabolic flight path, with the wings closed, for a short time. As it needs the same mean power as before, for a shorter (flapping) time, the work done by each gram of muscle while flapping also goes up; the muscles work more efficiently at the higher power. If you work out V_{mp} and V_{mr} for a bird that is bounding, the bird works at full power as it pulls up while flapping, which brings the muscles into their most efficient operating range. The acceleration that a small bird can pull while flapping is about 4g, meaning that it can maintain a "power ratio" (flapping time/cycle time) down to 0.25. This may push its maximum range speed up by 11%, or more. Bounding is a peculiarly bird habit, and seen only in smaller species that are specialised to fly in this way. The effects of bounding on air speed have been observed in the field [9], and are included in the *Flight* programme. A wind tunnel study would be interesting, but would require more space for vertical movements than is available in any wind tunnel that exists at the moment.

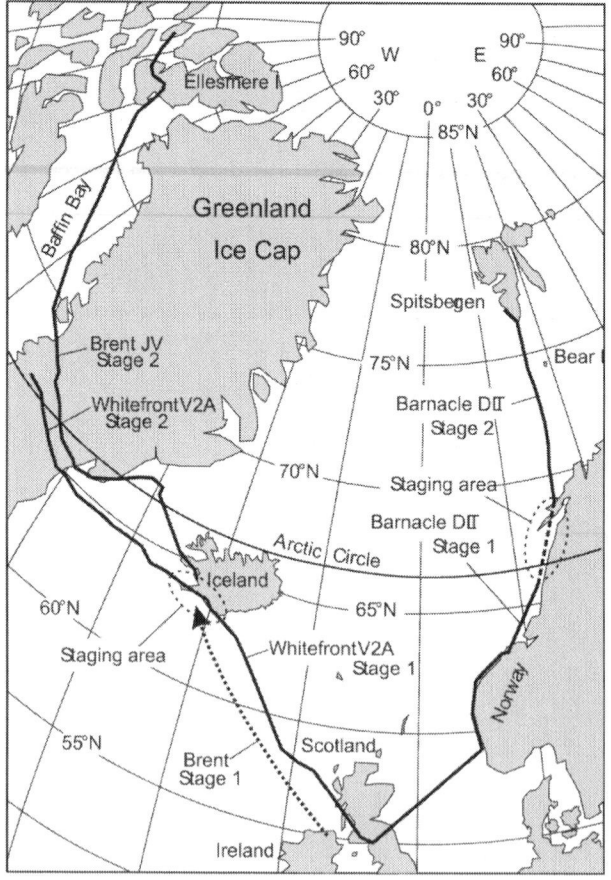

Figure 4. Migration routes of the three species of geese tracked during the spring migration of 2008. Positions were GPS recorded, usually every 2 h.

4. ENERGY HEIGHT FOR PLANES AND BIRDS

All of this depends of calculating the bird's fuel reserves in terms of its energy height, which is the height that the bird would get to, if all of its stored fuel were converted into height, using the engine (muscle) that it has. Figure 5 shows a bird starting from a known energy height. When the goose lands, the energy height continues going down (because of basal metabolism), and when the goose is moving, the energy height goes down a slope that is equal to its current lift:drag ratio. If our geese flew at their maximum range speed, they would come down on a gradient of about 19, but taking account of stops, they managed about 15 (Figure 6). No goose wants to arrive anywhere in spring below an energy height of 200 km, and those with longer ranges maintain 300 or 350 km. If the goose stops and feeds (and we knew where they did that), the energy height climbs up with little forward movement.

You work out the energy height from the current value of the fat fraction, which is the mass of consumable fat that the bird has on board, divided by the total mass of the bird (fuel and all). All birds with the same fat fraction and the same physiology come out at the same energy height, which is a logarithmic function of the fat fraction. The theory [10] says that the range is just the energy height multiplied by the lift:drag ratio. It would be simple to display energy height in a glass cockpit with an "energy altimeter", a fact which has not escaped the engineers [11]. Any pilot can multiply this by the lift:drag ratio, and any bird watcher can do it from the *Flight* programme, which displays both the energy height and the L/D ratio.

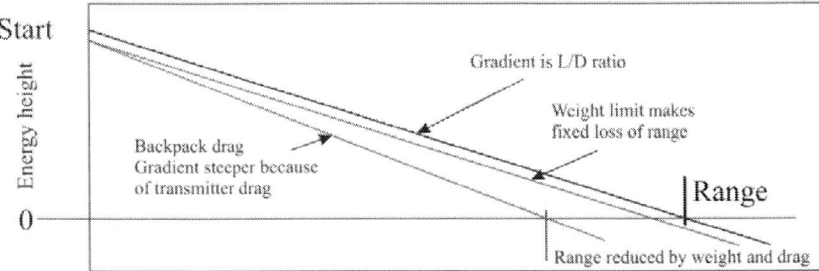

Figure 5. "Energy height" calculated from the fat fraction, represents the bird's fuel ratio, but does not reflect the bird's size. The "descent" from energy height in level flight is the same as the bird's lift-to-drag ratio. Extra drag uses up fuel continuously during a long flight, while the weight of the box may limit the amount of fuel that can be taken up initially, but has little other effect. The increased drag due to an added box (red line) consumes fuel whenever the bird is flying.

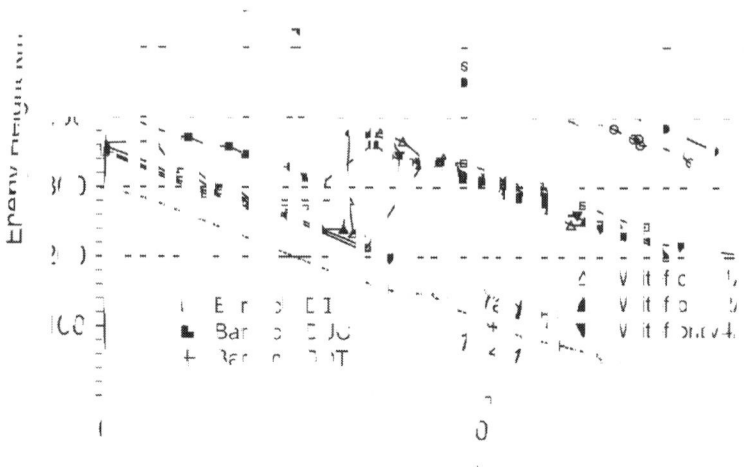

Figure 6. The energy heights achieved by different goose species are about 15:1, taking account of stops, as compared to a calculated maximum of about 19:1. Energy height is built up during feeding stopovers, and no goose wants to arrive anywhere below 200 km in spring. The Brent Geese, with the longest routes, stayed above 300 km. The fuel is not allowed to run low in spring migrations, but may do so in the autumn migrations.

Bird flight ranges are a little more complicated than those of aircraft, as all birds (it seems) have a dual fuel system in which the main energy supply source is fat. This is the best long-range fuel, because it can be stored in a non-hydrated form, which makes its energy density (3.9 MJ/kg) higher than for other forms of animal fuel. However, animals can also get usable energy by burning carbohydrate or protein, and a migrating bird, once it has been going for a few hours, no longer needs the amount of flight muscle that it needed to get airborne at the start [12]. That muscle becomes progressively surplus to requirements, so the bird burns it. Birds get most of the way by burning fat, and a lot further by burning their engines. The*Flight* programme takes account of this by taking the use of protein into account. This makes only a small difference on short flights, but a big difference to the range on longer routes, where the low energy density of hydrated protein means that burning a small amount of it gets rid of a lot of weight. Those godwits that migrate from Alaska to New Zealand started with a fat fraction of 0.55, and have enough fuel left to carry on to the South Pole, if they really wanted to do that.

5. FEATHERED WINGS

5.1. The Feathered Surface

The bird skin structures (feathers) are often said to be just another variant of a modified reptile scale, and that may be so, but they are made of different protein from other reptile scales, and have some special properties of their own. Feathers are the only known skin covering structures that contribute directly to the structure of the animal, and they have nothing to do with the vaguely fibrous "feather-like" structures seen in fossils that are later in time than the earliest birds (lower Cretaceous). The shafts of a bird's wing feathers form a distributive spar, which collects up the forces caused by the air flow on each feather, and sends it through the feather's root to the bony skeleton (Figure 7). The bird wing is a "distributed spar wing", not a "tension wing" as seen in bats and pterosaurs. The feathers not only form the structure of the wing, they determine its aerodynamics as well. The unusual shape of feathers is so distinctive that the first known bird feather from the early-bird fossil *Archaeopteryx* was known as "the feather" in the mid nineteenth century, and identified as coming from a bird, long before any other parts of these fossils were known [13]. It has a central rhachis (shaft), which is shaped to collect up the bending and torsional moments developed by the vanes, and deliver them through the root of each feather, to the bony skeleton of the arm. If a feather is lost, there is a gap in the wing, but the wing as whole continues to work, unlike a bat or pterosaur wing, which collapses if a part of the bony structure is lost. The whole structure of the bird wing, made of feathers, can be shed by molting, once per year or more often, and replaced.

5.2. Lowson's Contribution

It has been noted above that a feather-covered non-lifting bird body behaves in a way of its own, and that its lack of drag seems to be something to do with the feathered covering. The late Prof. Martin Lowson, formerly head of Aero Engineering at Bristol, studied the lifting of feathers in reversed flow, both on wings and bird bodies, but had trouble with referees in biological journals, and had not published this at the time of his death in 2013. He produced both theoretical and observational grounds that this lifting can be controlled by muscles under the skin, and can be used to defer or reverse flow separation, both on lifting and body surfaces. The Lund wind tunnel has a PIV system devised and installed by Geoff Spedding of the University of Southern California [14], whose primary purpose is to map the wake velocities, downstream of a small bird in flapping flight. Observers trying to observe separated flow with this device easily detected it in wing models, but had great difficulty in seeing any such effects behind a flapping wing of a live bird. The feathered surface clearly has something to do with these effects, and the PIV system might be able to find out how it works at the level of individual feathers, with respects to Prof. Lowson.

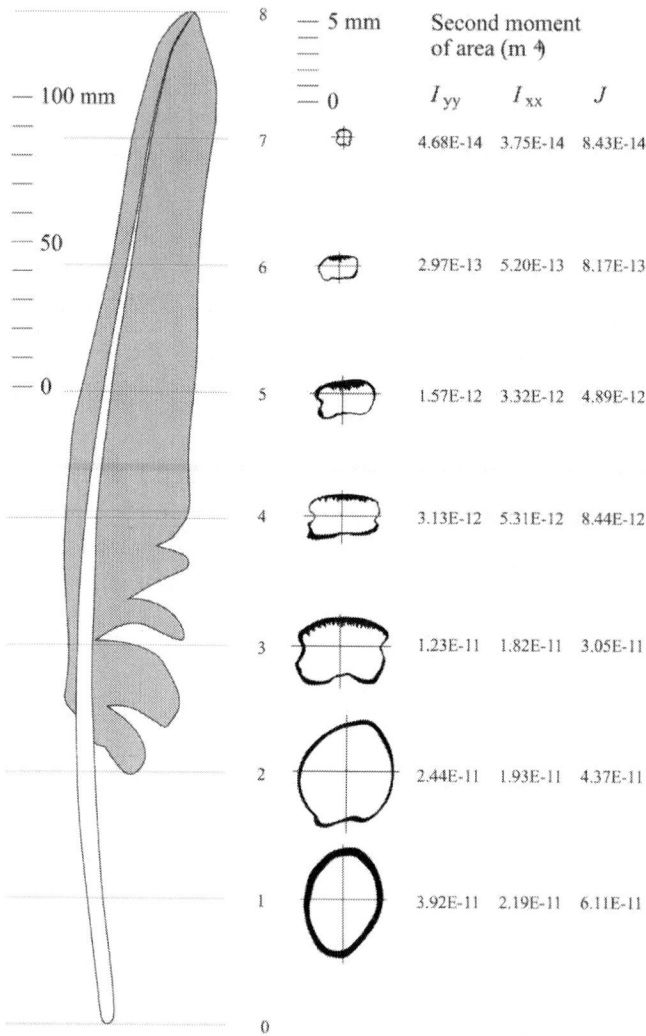

		Second moment of area (m⁴)		
		I_{yy}	I_{xx}	J
7		4.68E-14	3.75E-14	8.43E-14
6		2.97E-13	5.20E-13	8.17E-13
5		1.57E-12	3.32E-12	4.89E-12
4		3.13E-12	5.31E-12	8.44E-12
3		1.23E-11	1.82E-11	3.05E-11
2		2.44E-11	1.93E-11	4.37E-11
1		3.92E-11	2.19E-11	6.11E-11

Figure 7. The shaft of a Greylag Goose secondary flight feather is made of keratin, and has a second moment of area that carries the bending and torsional loads of that feather (not of other feathers). The late-Jurassic *Archaeopteryx* fossils had wing feathers exactly like this, showing that the bird wing was developed in Jurassic times. This was about 110 million years before the extinction of the pterosaurs, which had no such structures.

6. TENSION WINGS

6.1. The Tension Wing of Bats

There is another type of modern wing, that of bats, which started from nothing after the loss of the dinosaurs, and developed alongside the post-dinosaur birds. Fruit bats were flown in Nairobi with the original blower tunnel (Figure 8) [15,16]. It allowed 3-dimensional stereo pictures of the wing membranes to be obtained, which were good enough for rough lift and drag analysis, and showed gliding performance much like a pigeon. Several species of small insectivorous and nectar-eating bats have been flown in the Lund wind tunnel, mainly to observe their wakes with the PIV installation [17]. Bat wings are held in shape by a skeletal framework, stretching a membrane that has no stiffness of its own, and is not stiff in itself, as bird wings are. In flight the planform has only a limited degree of variability of shape, because relaxing the muscles that hold the framework in shape also takes the tension out of the main membrane. Bats have a wholly different system of muscle fibres in the membrane, which tighten up to reduce its curvature, without connecting with the skeleton. Their wings are exceptionally good at low-speed manoeuvring flight, but are not general purpose wings, adaptable like those of birds for swimming, running on the ground, catching other birds in the air and so on. Birds use every kind of habitat known to mammals, whereas bats are fruit and honey eaters, and (especially) catchers of slow-flying insect prey. Bat eyes do not dominate their skull, and they are mostly nocturnal animals with senses that are based around hearing. Bats go up to the size of a small goose, and do not reach the size of swans, or do the long-distance migrations that are so familiar in birds.

6.2. The Tension Wing of Pterosaurs

It would be an error to think of birds as being permanently the best flying animals, although they have held that position for some millions of years. Their modern time began in the Palaeocene, when a few birds survived from the disaster that finished the dinosaurs, probably an impact with a small asteroid about 65 million years ago. Their rivals, the pterosaurs [18] disappeared from the fauna, along with all the remaining dinosaurs, never to return. The period from about 250 million years ago up till 65 million years ago followed the great extinction which ended the preceding Permian period. At some time in the Triassic period, which followed the great Permian extinction, both the birds and the pterosaurs started from scratch, as non-flying arboreal creatures, living in the trees of the time. Their anatomy is radically different from the start, showing the contrast between the multiple-spar wing, built around feathers (above), and the tension wing seen in the pterosaurs. The pterosaur wing [19] is made up from a flexible protein membrane, which (unlike feathers) has no strength to resist the bending and torsional moments developed by a wing. The resistance to these forces comes not from the wing area itself, but from the bony framework surrounding it, made up from the side of the body and the arm skeleton, which had just one extremely elongated wing finger along the leading edge. The wing was held in shape by the muscles controlling the body framework, and when these muscles relaxed (as in death) the wing simply collapsed. It is like a hang glider wing, but not like a bird wing, which does not collapse when the bird dies. Figure 9 is a reconstruction of an early pterosaur wing from about 150 million years ago. It had an elastic membrane which was pulled into shape by the muscles, and required the side of the body to be suitably shaped to hold it in position, whereas the bird wing shape does not depend on the shape of the body or legs. Figure 9 shows some differences from the common reconstruction of pterosaurs.

The astonishing thing about this is that the very same German Jurassic rocks, from Solnhofen in Bavaria, that held the fossils of *Archaeopteryx*, the first known bird, also held a good selection of pterosaurs from the same time and habitat. This was not the end of the pterosaur time, but the middle, and the really big pterosaurs were then still millions of years in the future. How, then did birds and pterosaurs carry on side by side throughout the Cretaceous period of time, another 60 or so million years, which ended with the destruction of the dinosaurs and pterosaurs? Why did not birds, with much better wings, replace the pterosaurs? In my view there is a reason for that, and it is to do with the automatic homing system that bird that modern birds (the few survivors) have built into their brains. They have a time sense that will keep track of time within seconds for months at a time, in the absence of external stimuli, and the ability to do trigonometry relating a currently observed plane (the horizon) to a world-sized sphere. That is all it takes to navigate anywhere on earth [20]. It is not unlikely that just one group of birds could do that at the time of the loss of dinosaurs, and that they were the only ones that were able to survive, and become the ancestors of all modern birds.

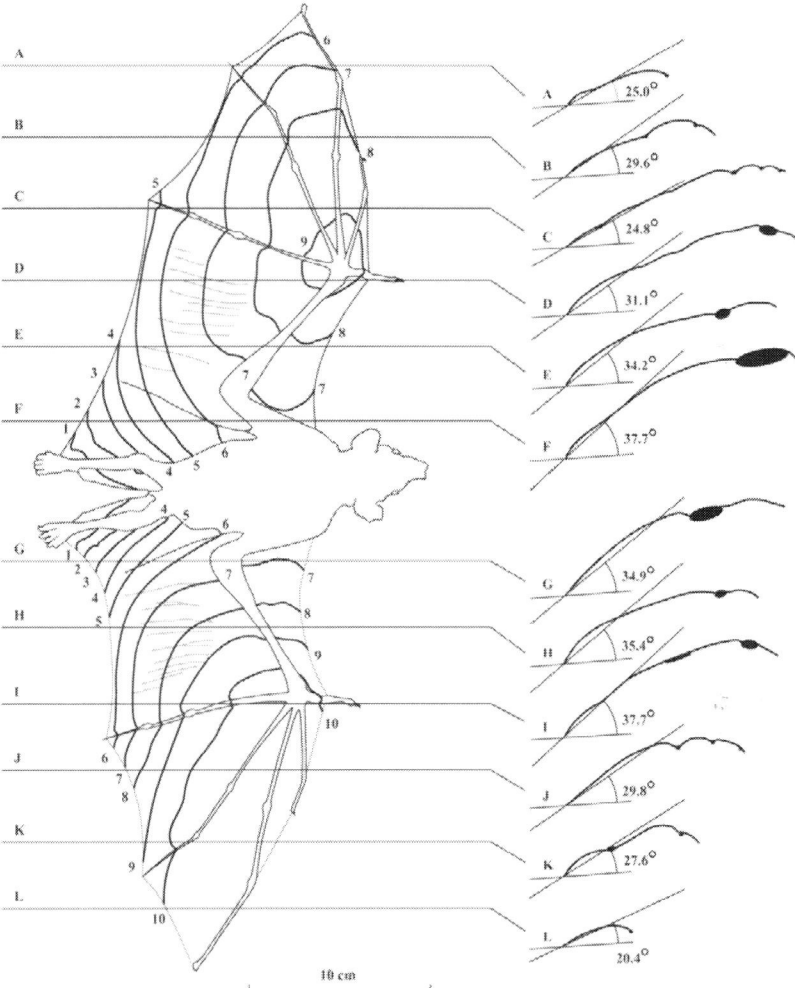

10 cm

Figure 8. A *Rousettus* fruit bat gliding at Nairobi in the original Bristol blower tunnel, illuminated by a flashgun shining upwards through the wing. This is one frame of a stereo pair. All the strength comes from the bones of the skeleton, not from the membrane itself. The bat cannot adjust its wing in the manner of Figure 3, as this would unload the membrane. The thin muscles in the membrane, which do not attach to the skeleton, are used to flatten the wing profile at higher speeds. The sections are all convex at low speed, but the bat flew perfectly steadily, with no signs of instability.

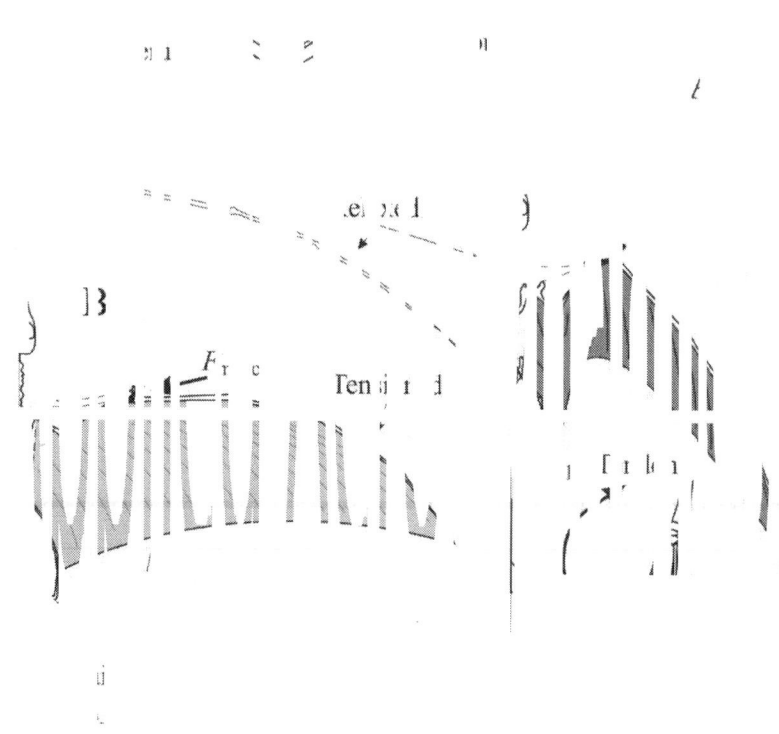

Figure 9. Reconstruction of a blackbird-sized pterosaur. (**A**) Wing skeleton without membrane; (**B**) The wing is pulled open by the wrist muscle (F_{musc}), and elastic fibres in the membrane maintain tension as it is stretched; (**C**) The surface wrinkles ("stiffening fibres") are surface features that only appear when the membrane contracts, especially when it is dead; (**D**) The elongated 5th toe stretches the trailing-edge tendon.

7. WHY STUDY BIRDS?

Ornithologists, like biologists of any kind, want to know where birds came from, and (nowadays) what the potential of flying animals might be, on planets that differ from our own home, here on Earth. We have, effectively, one good wind tunnel in Lund in which to continue this study, and another at Seewiesen that is missing some features. We have a theory that predicts in some detail what migrating birds can do, and how their bodies work in long-range flights like those of the BBC geese [8]. Despite the evidence presented, current bird watchers seem reluctant to believe that it really is possible to estimate the physical details of a migration flight. That may be because we do not have direct observations of real birds, to see whether their behavior is as predicted by the theory. The facilities to test this were built into the Lund wind tunnel, which can simulate long migratory flights. The bird can set the flight speed, and the tunnel can be stopped at any time to weigh the bird, and check its body composition. The theory predicts in detail what the bird should do in these circumstances, and checking this is a matter of training and prolonged experiments, to see whether the theory is basically correct, and amenable to fit what birds actually do. Bird watchers assume that the results in [8] are from a theory about geese, but they are not. The *Flight* programme applies to any birds whose wing span, wing area and mass are known, flying in air of known density, with a known strength of gravity. Testing it on any bird, in any wind tunnel, will test it for all.

I say [7] that modern birds are not like other animals. The birds that survived the dinosaur crash are the only living animals with a fully automatic, long period, spherical celestial navigator that works without modification anywhere in the world. Disprove it if you can.

ACKNOWLEDGMENTS

My views come from watching birds and from arguments with pilots and birdwatchers alike. It is a pleasure to thank them all.

REFERENCES

1. Pennycuick, C.J. A wind tunnel study of gliding flight in the pigeon *Columba livia*. *J. Exp. Biol.* **1968**, *49*, 509–526.

2. Pennycuick, C.J. *Modelling the Flying Bird*; Elsevier: Amsterdam, The Netherlands, 2008.

3. Hill, A.V. The dimensions of animals and their muscular dynamics. *Sci. Prog.* **1950**, *38*, 209–230.

4. Pennycuick, C.J.; Alerstam, T.; Hedenström, A. A new wind tunnel for bird flight experiments at Lund University, Sweden. *J. Exp. Biol.* **1997**, *200*, 1441–1449.

5. Pennycuick, C.J. Power requirements for horizontal flight in the pigeon *Columba livia*. *J. Exp. Biol.* **1968**, *49*, 527–555.

6. Pennycuick, C.J.; Fast, P.L.F.; Ballerstädt, N.; Rattenborg, N. The effect of an external transmitter on the drag coefficient of a bird's body, and hence on migration range, and energy reserves after migration. *J. Ornithol.* **2011**, *153*, 633–644.

7. Pennycuick, C.J.; Pennycuick, S. *Birds Never Get Lost*; Troubador: Leicester, UK, 2015.

8. Pennycuick, C.J.; Griffin, L.R.; Colhoun, K.; Angwin, R. A trial of a non-statistical computer program for monitoring fuel reserves, response to wind and other details from GPS tracks of migrating geese. *J. Ornithol.* **2011**, *152*, 87–99.

9. Pennycuick, C.J. Speeds and wingbeat frequencies of migrating birds compared with calculated benchmarks. *J. Exp. Biol.* **2001**, *204*, 3283–3294.

10. Pennycuick, C.J. The concept of energy height in animal locomotion: Separating mechanics from physiology. *J. Theor. Biol.* **2003**, *224*, 189–203.

11. Merkt, J.R. Flight energy management training: Promoting safety and efficiency. *J. Aviat. Technol. Eng.* **2013**, *3*, 24–36.

12. Piersma, T.; Gill, R.E. Guts don't fly: Small digestive organs in obese bar-tailed godwits. *Auk* **1998**, *115*, 196–203.

13. Wellnhofer, P. *Archaeopteryx the Icon of Evolution*; Pfeil: Munich, Germany, 2009.

14. Hedenström, A.; Rosen, M.; Spedding, G.R. Vortex wakes generated by robins *Erithacus rubecula* during free flight in a wind tunnel. *J. R. Soc. Interface* **2006**, *3*, 263–276.

15. Pennycuick, C.J. Gliding flight of the dog-faced bat *Rousettus aegyptiacus*, observed in a wind tunnel. *J. Exp. Biol.* **1971**,*55*, 833–845.

16. Pennycuick, C.J. Wing profile shape in a fruit bat gliding in a wind tunnel, determined by photogrammetry. *Period. Biol.* **1973**, *75*, 77–82.

17. Hedenström, A.; Muijres, F.T.; von Busse, R.; Johansson, L.C.; Winter, Y.; Spedding, G.R. High-speed stereo DPIV measurement of wakes of two bat species flying freely in a wind tunnel. *Exp. Fluids* **2009**, *46*, 923–932.

18. Wellnhofer, P. *The Illustrated Encyclopedia of Pterosaurs*; Salamander: London, UK, 1991.

19. Pennycuick, C.J. On the reconstruction of pterosaurs and their manner of flight, with notes on vortex wakes. *Biol. Rev.***1988**, *63*, 299–331.

20. Pennycuick, C.J. The physical basis of astro-navigation in birds: Theoretical considerations. *J. Exp. Biol.* **1960**, *37*, 573–593.

CHAPTER 10

Numerical investigation of flow separation behavior in an over-expanded annular conical aerospike nozzle

Miaosheng He , Lizi Qin·, Yu Liu

School of Astronautics, Beihang University, Beijing 100191, China.

ABSTRACT

A three-part numerical investigation has been conducted in order to identify the flow separation behavior—the progression of the shock structure, the flow separation pattern with an increase in the nozzle pressure ratio (NPR), the prediction of the separation data on the nozzle wall, and the influence of the gas density effect on the flow separation behavior are included. The computational results reveal that the annular conical aerospike nozzle is dominated by shock/shock and shock/boundary layer interactions at all calculated NPRs, and the shock physics and associated flow separation behavior are quite complex. An abnormal flow separation behavior as well as a transition process from no flow separation at highly over-expanded conditions to a restricted shock separation and finally to a free shock separation even at the deign condition can be observed. The complex shock physics has further influence on the separation data on both the spike and cowl walls, and separation criteria suggested by literatures developed from separation data in conical or bell-type rocket nozzles fail at the prediction of flow separation behavior in the present asymmetric supersonic nozzle. Correlation of flow separation with the gas density is distinct for highly over-expanded conditions. Decreasing the gas density or reducing mass flow results in a smaller adverse pressure gradient across the separation shock or a weaker shock system, and this is strongly coupled with the flow separation behavior. The computational results agree well with the experimental data in both shock physics and static wall pressure distribution at the specific NPRs, indicating that the computational methodology here is advisable to accurately predict the flow physics.

KEYWORDS

Aerodynamics; Aerospike nozzle; Flow simulation; Flow separation; Gas density effect; Over-expanded flow; Reynolds-averaged Navier–Stokes (RANS)

1. INTRODUCTION

The flow separation in supersonic convergent–divergent nozzles is a basic fluid-dynamics phenomenon that occurs at a certain nozzle pressure ratio (NPR), resulting in the presence of shock waves and shock/boundary layer interactions inside nozzles. It has been the subject of various experimental and numerical studies in the past. Today, with the renewed interest in supersonic flights and space vehicles, the subject has become increasingly important, especially for aerospace applications for rockets, missiles, supersonic aircraft, etc. There has been a widespread desire to investigate features with shock/boundary layer interactions in highly over-expanded rocket nozzles, since these interactions are responsible for acoustic, vibrate-acoustic, thermal, and mechanical-induced loads that act on the structure. Conventional bell-type nozzles suffer from the above interactions with reduced engine performance at low-altitude highly over-expanded conditions due to fixed geometries. Based on the above background, different types of nozzle concept with altitude-adapting capabilities have been developed and tested on the ground in the past, and the aerospike nozzle is included as a strong contender for the propulsion system of reusable spacecraft.1 and 2 In the recent years, a renewed interest in the aerospike nozzle flowfield has been generated for both rocket and aeronautic applications. 2·3·4·5 and 6 However, while flow separation in over-expanded planar or bell ideal and optimized contour nozzles has been widely investigated to elucidate the phenomenon of boundary layer separation and shock interactions, aerospike nozzles have received little attention in the frame of this study. Verma[4] and Kapilavai et al.[7] presented pioneering work upon flow separation behavior in aerospike nozzles that operated at NPRs below 10, and both of the two studies indicated that the presence of flow phenomenon was associated with nozzle flow separation, and unsteady shock oscillation induced by the interaction of the shock/boundary layer seen in conventional supersonic nozzles with diverging sections could also be expected in aerospike nozzles at off-design operating NPRs. In spite of few rare studies on this subject, understanding of the flow separation behavior as well as fundamental knowledge of supersonic flow physics in the presence of shock wave propagation, shock reflection at walls, and shock/shock and shock/boundary layer interactions in such a convergent-divergent nozzle are still needed.

Studies of flow separation in supersonic nozzles dated back to the 1960s with the first work of Arens and Spiegler[8], who published the first approach to include the Mach number influence in the theoretical prediction of free shock separation. Schmucker[9] continued the research through the later years in the 1970s, and based on a simplified boundary layer integral approach, Schmucker proposed the famous purely empirical criterion for free shock separation (FSS) which is still widely used to date. From several experimental studies, performed on either full-scale[10] or subscale[11, 12 and 13] optimized nozzles, and corroborated by different numerical simulations[14, 15, 16, 17 and 18], the presence of two distinct flow separation patterns, namely FSS and restricted shock separation (RSS), is demonstrated. A transition in the separation pattern from FSS to RSS and vice-versa might occur, which was firstly observed in the early 1970s by Nave and Coffey.[10] At the initial state of start-up or when a supersonic nozzle operates at low NPRs, the flow mostly resides in an FSS state[19], as shown in Fig. 1.[20] A single incipient separation of the flow along the interior surface of the nozzle is triggered by an adverse pressure gradient between the regions of isentropic expansion and subsonic entrainment. The shock that originates from the incipient separation line interacts with a reflected shock; this shock emanates from the triple-point which is the location where the Mach disk, internal and reflected shocks coincide. In the FSS state, a separation region forms which encompasses a series of compression/expansion waves, and this separated flow fails to reattach back to the wall at low NPRs due to the lack of outward radial momentum as a free supersonic annular jet. A recirculating subsonic region forms between the separated free annular jet and the nozzle wall, which entrains ambient air along the nozzle wall and adapts the static wall pressure to the ambient condition.

The RSS state refers to the canonical shock/boundary layer interaction which is present in many high-speed devices. For example, RSS is known to occur in thrust optimized parabola nozzles of engines like Vulcain, space shuttle main engine (SSME), or J-2S during the start-up process. RSS is characterized by a small separation region or bubble which exists immediately downstream from the shock wave, in which the mean flow circulates and separates or tilts away from the wall before the flow reattaches and continues down the length of the nozzle as an attached boundary layer. Depending on the nozzle contour and expansion ratio, more than one bubble may be present, as shown in Fig. 2.[20] Upon further increases in the NPR, the RSS flow regime will translate downstream and the enclosed separation bubble opens up to the ambient environment when passing over the nozzle lip. When the last annular separation bubble downstream from the incipient separation shock opens up, the flow structure then switches to an FSS state due to the presence of a single separation shock with an associated separated flow region downstream.

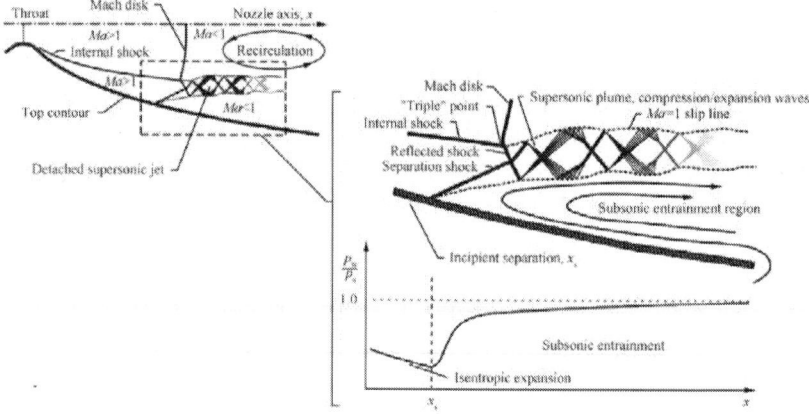

Figure. 1. Illustration of the internal shock structure in a thrust optimized parabolic nozzle during an FSS state.[20]

In the past, many researchers from different groups in Europe21 [and] 22, USA[10], and Japan[23] have distinguished these flow separation behaviors in rocket nozzles and have identified a transition phenomenon between the two separation regimes during the start-up or shut-down process. In a typical rocket engine, the combustion chamber pressure rises from the ambient pressure to the steady-state operating condition14 17 [and] 24, and flow separation occurs when the chamber pressure is relatively low, so as to yield a static wall pressure much lower than the ambient one in some locations of the nozzle divergent. During the start-up process, the nozzle flow is popularly dominated by the FSS structure, and then when the combustion chamber pressure rises over a certain critical value, the FSS regime is replaced by the RSS regime. The identification of this transition is important for nozzles applications in rocket engines or supersonic aircraft, as it is directly attributed to the largest lateral loads seen during their operations.16 [and] 25 Previous studies on this subject have also shown that these flow features in the presence of FSS/RSS transition, unsteady shock motion behavior, and side loads are strongly dependent on the nozzle geometry, the strength of the shock, and the over-expanded ratio. A recent paper by Ostlund and Muhammad-Klingmann[26] reviewed several conditions and regimes which led to the most severe side-load[10], and commented on the difficulty in accurately modeling these phenomena.

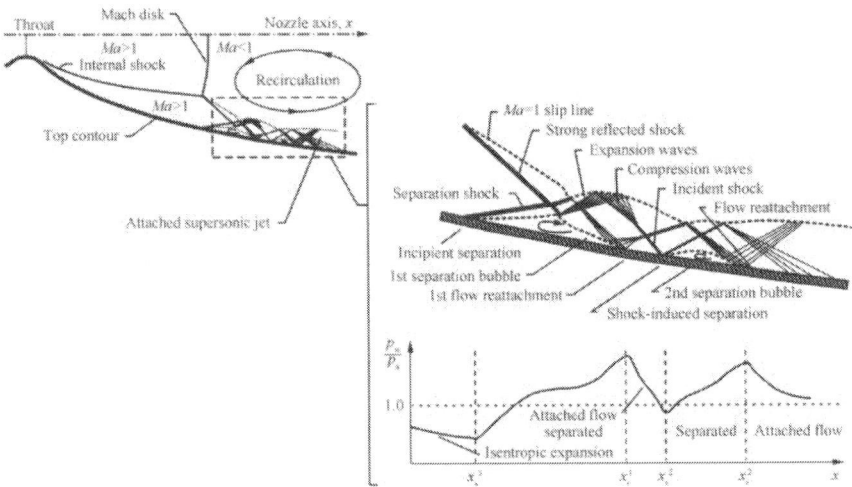

Figure. 2. Illustration of the internal shock structure in a thrust optimized parabolic nozzle during an RSS state.[20]

In the case of aerospike nozzles, although they have been extensively studied experimentally and numerically since the 1950s[27, 28, 29, 30 and 31], particular attention has been mostly paid to the so-called altitude-compensating capacity.[15, 16 and 28] A considerable body of theoretical and experimental data indicates that this altitude-compensating capacity enables an aerospike nozzle to eliminate over-expansion losses altitudes below the design point, and to provide a gain in performance relative to a conventional convergent-divergent supersonic nozzle.[5] To get the highest benefit with this nozzle concept, the design pressure ratio or the design geometrical area ratio is always chosen as high as possible. Therefore, for rocket or supersonic applications, the lower operating NPRs during the start-up process can also result in the shock structure to reside within the internal nozzle exit or on the spike surface, and the presence of flow phenomenon associated with nozzle flow separation seen in supersonic nozzles with diverging sections can be expected in aerospike nozzles.

Verma et al.4 32 and 33 presented a series of experimental studies of shock physics and nozzle performance in two classic aerospike nozzle concepts, and the conical annular and linear with low-angle plug configurations, truncated and full length with or without freestream effects were also considered. For an over-expanded 15° annular conical aerospike nozzle[4], it was observed that the over-expansion shock from the internal nozzle, the over-expansion shock on the spike surface, and the expansion fan from the cowl lip of the internal nozzle dominated the overall flowfield development. Increasing the NPR changes the angles of these shocks as the internal nozzle operates from the over-expanded to the under-expanded condition. This produces three different types of flow separation conditions on the spike. The paper gave an idea of the flowfield once the shock moved out of the internal nozzle, while the flow features when the shock structure resided within the internal nozzle section at lower NPRs were not available. However, complex shock physics and flow features can be expected in this spectrum of NPRs as indicated by present numerical studies. Kapilavai et al.[7] dedicated his study to the aerodynamic characteristics of a so-called shrouded plug nozzle at various operating conditions, and the nozzle was also observed to be dominated by shock/shock and shock/boundary layer interactions at all off-design NPRs. Depending on shock interaction with the boundary layer on the plug wall, the nozzle exhibited both fully separated and reattached boundary layer regimes. However, it should be noted that the shrouded plug nozzle whose shroud or cowl extends over a considerable portion offers both aerodynamic and structural characteristics with respective to a more conventional aerospike nozzle. These results and comments indicate that a coupled research effort between experiment and numerical simulation is needed to fully understand the shock physics in aerospike nozzles for the complete spectrum of nozzle pressure ratios.

The current work attempts to reproduce and to expand the experimental studies of Verma.[4] The aim of the paper is to numerically study in detail the flow separation behavior at sea-level as well as imposed high-altitude simulation conditions without freestream flow, for the purpose of providing an insightful understanding of the shock physics and characteristics of shock/shock and shock/boundary layer interactions at various operating conditions, from highly over-expanded conditions to the designed point of the nozzle, in particular, the shock-induced flow separation behavior and its evolution process with the change of NPR. A comparison of the separation data against a series of separation criterions form literatures is conducted to address their applicability for the prediction of flow separation behavior in such an asymmetric supersonic nozzle. Additionally, a comparison of flow separation behavior at sea-level conditions against the imposed high-altitude simulation conditions is to demonstrate the gas density effect on afore-mentioned flow characteristics in present aerospike nozzles; in particular, its effect on the flow separation behavior in highly over-expanded conditions has been studied when more complex shock/boundary layer interactions can be expected.

2. NUMERICAL ANALYSIS

2.1. Experimental details and modeling

The experimental work performed on the present aerospike nozzle was carried out in a 0.5-m base flow facility, a special purpose blow-down-type tunnel, detailed in Ref.[4]. The salient features relevant to this study are shown in Fig. 3, in which the 15° half-angle θ annular conical aerospike nozzle model with a design Mach number of 2.0 was mounted on the central cylindrical inner body. The afterbody contour was designed based on the recommendations given in Ref. [34]. To obtain the steady pressure distribution on the spike, up to 12 pressure points with a pitch of 4.0 mm were installed at axial locations along a single line on the spike surface. The nozzle exit radius r_e is 25 mm, the annular gap at the throat section h_t is 9 mm, and the length of the spike or plug L is 59.71 mm. The aerospike nozzle area ratio is defined as the ratio of the area at the spike end A_e to the annular throat area A_t. The length of the cowl, measured as the distance from the throat section to the cowl lip, l is fixed as 9.0 mm, resulting in an area ratio of the inner nozzleε_i = 1.19.

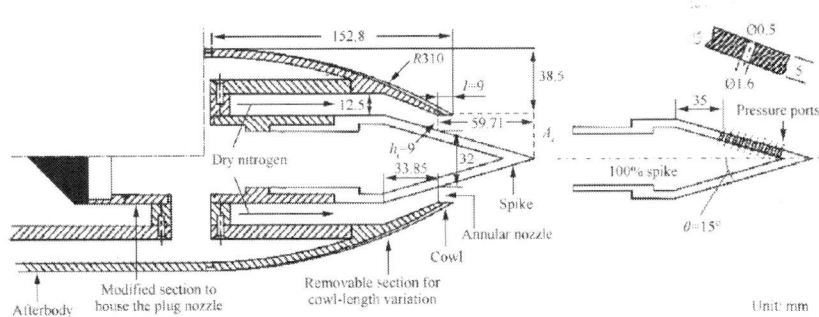

Figure. 3. Schematic of the experimental annular conical aerospike nozzle model.[4]

In the present case, a modified experimental annular conical aerospike nozzle model is used, resulting in an annular axisymmetric nozzle configuration, as seen in Fig. 3, where the flow phenomena of primary interests takes place. The modified section to house the plug nozzle is not included in the axisymmetric computations; this strut is unlikely to have major effect on the nozzle flowfield as it is located in the upstream subsonic convergent section of the nozzle. A straight annular tube similar to the one in the experiments is utilized in the present numerical model to provide an area-constant expansion region until the subsonic streams reach the beginning of the nozzle convergence.

2.2. Flow conditions

The computational flow conditions reproduce and extend on the experimental conditions, that is, NPR covers a wide range of 1.60–9.87, corresponding to the unchoked subsonic condition to a slightly under-expanded condition. In the experiments, NPRs were conducted as $2.10 \leqslant NPR \leqslant 5.75$ for the full-length spike analysis. As indicated in Section2.1, the presence of the cowl results in an additional diverging section of the shock interacting with the boundary layer on the upstream spike and cowl surface at low NPRs, where the main flow phenomena were unable to be captured by the schlieren images in the experiments. The present NPRs design will provide a complete flow expansion spectrum for such a supersonic nozzle configuration.

Three environment back pressure, p_b, or ambient pressure, p_a, conditions are designed to conduct the gas density effect analysis, as shown in Table 1. In Table 1, p_{ON} is the total pressure at the nozzle inlet. Comparisons are made between a sea-level case and two imposed high-altitude simulation cases with back pressure values of 50% and 25% of the sea-level ambient pressure condition, respectively. The NPRs conducted are identical for all three cases. The objective of the simulation is to assess the impact of the low ambient pressure environment, resulting in low nozzle total pressure and then low gas density for the NPR mimic on the shock structures and flow separation characteristics. In the present study, the Reynolds number is based on the hydraulic diameter defined as the incipient separation location ranged from 1.59×10^6 to 34.93×10^6, corresponding to a fully turbulent boundary layer.

2.3. Numerical procedure

The numerical study has been conducted using a finite-volume unsteady Reynolds-averaged Navier–Stokes (RANS) solver. The two-equation shear stress transport (SST) model of Menter et al.[35] is used here to describe the turbulence. Previous successful efforts to compute the internal nozzle separated flow using RANS-type models by Hunter[36], Xiao et al.[37, 38 and 39], and Carlson[40] indicate that the SST model provides a reasonable prediction of flow separation just downstream from the shock caused by the shock/boundary layer interaction inside the nozzle, and therefore the best capture of the shock location and pressure distribution against the experiment. In present study, all computations are made using the axisymmetric assumption with a primary motive of a fast and efficient means of obtaining insight into the relevant shock structure and flow separation behavior at various operating conditions. Later, when the computational results are compared with the experimental data, it will be shown that the axisymmetric assumption is accurate for the prediction of salient shock physics as well as static wall pressure.

Table 1. Condition cases for gas density effect analysis.

Ambient pressure conditions	Case No.	NPR	p_{ON} (kPa)	p_a (Pa)	Mass flow rate(kg/s)
Sea-level atmosphere	1-1	1.60–9.87	162.12–1000.09	101325	0.4236–2.6129
50% of sea-level ambient pressure	1-2	1.60–9.87	81.06–500.04	50662.5	0.2118–1.3065
25% of sea-level ambient pressure	1-4	1.60–9.87	40.53–250.02	25331.25	0.1059–0.6521

Fig. 4 shows the nozzle numerical model, including the detailed grid distribution in the nozzle region where the flow phenomena of primary interest take place. The computational domain includes the domain inside the nozzle and an ambient region around the outer surface. The domain extends $15D$ (where D is nozzle exit diameter) upstream from the throat, more than $40D$ downstream, and $10D$ in the direction perpendicular to the axis. In terms of grid structure, multi-block structured grids have been used in this calculation and are shown in Fig. 4(b). Only every second grid line in each direction is displayed for clarity. For the nozzle region, grid density is higher in the divergent part of the nozzle to improve the resolution for capturing shocks. In addition, the grid is clustered along the cowl and spike walls to resolve the boundary layers, for a Reynolds number based on an order of 10^6, and the first grid point from either the spike or the cowl wall gives a $y^+ < 1$. A detailed discussion of the grid convergence study is reported in the next section.

Fixed total conditions have been employed for the outerface boundary, outlet upstream, outlet lateralface, and outlet downstream, the total pressure is set to the ambient pressure, and the boundary condition is designed in such a way that the supersonic plume may go out of the boundary, while the ambient air may come into the boundary due to the possible plume entrainment effect. The total pressure and the total temperature at the nozzle inlet are specified to be $p_{ON} = NPR \times p_a$ and $T_{ON} = T_a$, respectively. For all cases, the stagnation and the ambient temperature, T_{ON} and T_a, in the computations are fixed as 300 K, and pure gaseous nitrogen is simulated as the nozzle jet while a perfect gas with the properties of air is treated as the ambient fluid. A no-slip, adiabatic condition is specified for the solid walls, and a symmetric condition is imposed on the nozzle axisline. In addition, the upstream turbulence intensity is prescribed at the nozzle inlet boundary based on the located hydraulic diameter.

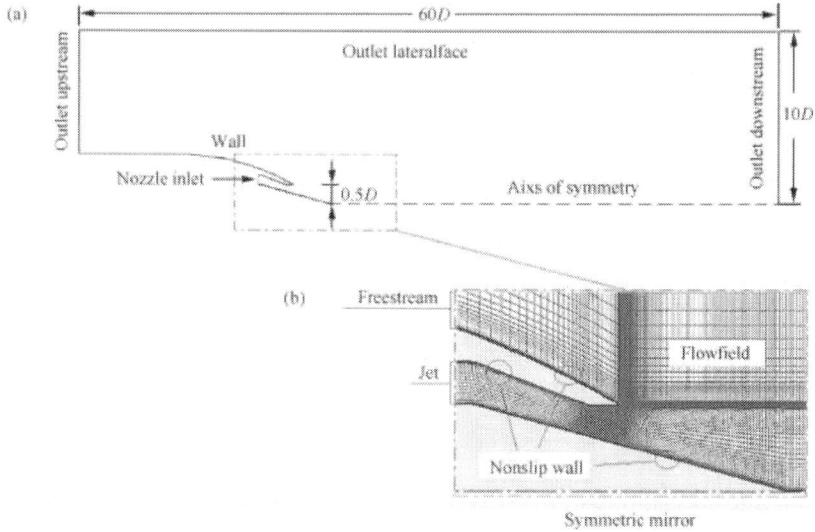

Figure. 4. Illustrations of the nozzle numerical model: (a) overall computational domain, both inside and outside the nozzle, and (b) a close-up view of mesh topology in vicinity of the nozzle region.

Preconditioning is employed for accelerating the calculation at low subsonic NPRs. However, the computed shock structures show the flow to assume a steady state in contrast to the experiment, in which a 1500-Hz peak in frequency can be seen in the region of separation.[4] This is particularly true when RANS-type time-averaged models are utilized to compute an internal nozzle flow with random unsteadiness or distinct acoustic tones. For the present study, all the computations are conducted with an RANS-type modeling method in the primary interest of salient shock structure and flow separation characteristics development as the NPR is increased from low to high values in such an asymmetric supersonic nozzle.

2.4. Grid convergence

In order to establish the fidelity of the numerical database, we have conducted a grid convergence study on the nozzle, and a series of three computational meshes has been run. As it is important to accurately capture the salient shock structures and their interaction with the boundary layer which drives the flow separation behavior. We begin the discussions by comparing the computational flow patterns and the static wall pressure profiles at specific NPRs with the data taken from experiments at similar NPRs for the grid convergence study. In the experiments, when increasing the NPR values, the internal nozzle operates from the over-expanded to the under-expanded condition, three types of flow separation patterns can be observed at NPR = 2.10, 2.57, 3.82, respectively.[4] The comparisons in this section are all for the above three NPR conditions. Schlieren images have been used in the experiments to help in identifying the first gradients of density, which can exhibit the shock patterns in the flowfield. To make a direct comparison of the computational flowfield with the experimental results, the computational results also use the numerical schlieren pictures contoured by the plots of the absolute values of the first gradients of density at the grid nodes.

The shock-induced flow separation patterns for various mesh resolutions are displayed in Fig. 5, in which the numerical schlieren pictures listed from the first to the third row show the results based on the coarse (Mesh A), medium (Mesh B), and fine grid (Mesh C) cases, respectively. For all the three meshes, the three types of flow separation patterns at specific NPRs are well reproduced, and the numerical results are in very good agreement since the shock structures are very close to each other.

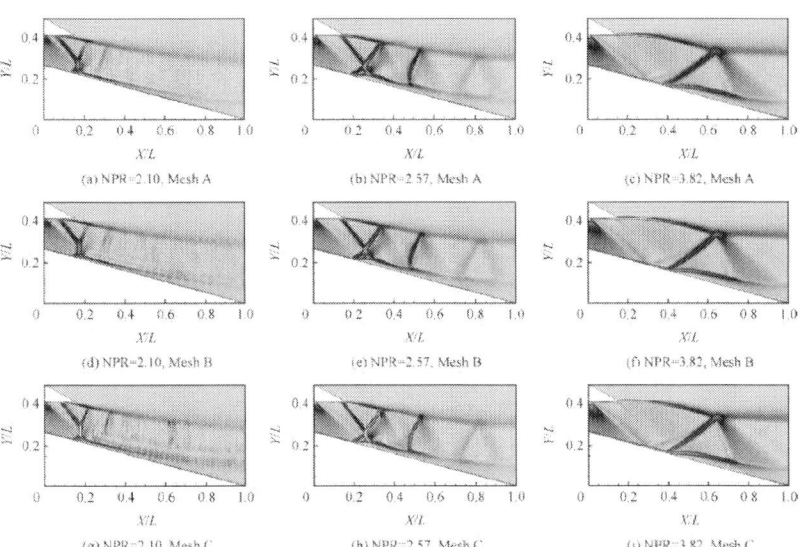

Figure. 5. Comparison of shock-induced flow separation patterns for various mesh resolutions.

The pressure distributions along the nozzle spike and the cowl wall have also been compared for the three meshes. Fig. 6 shows the static wall pressure p_w profiles normalized by the ambient pressure, p_a, at the three specific NPRs of 2.10, 2.57, and 3.82, respectively. Again, the plots are really similar for the three meshes, but a slight shift is observed between the coarse grid and the two finer grids. In addition, it is shown that the results for the medium and fine meshes collapse almost perfectly. It is noteworthy that the shock structures and location as well as the separation point are well predicted whatever mesh is considered. As a consequence, the so-called fine grid mesh is used in the present study, as the shock pattern captured is more elaborate as the grid is refined in the nozzle flow region, especially in highly over-expanded conditions when more complex shock/boundary layer interaction can be expected.

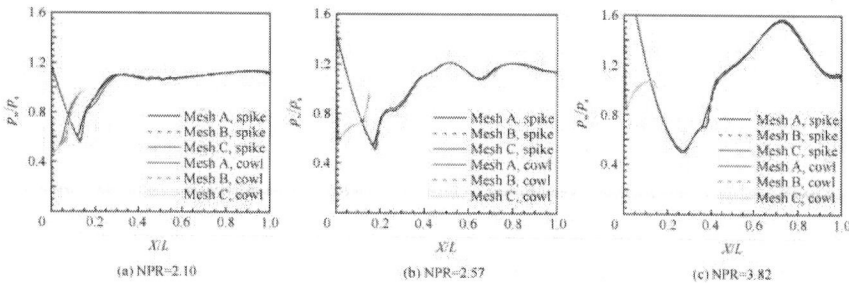

Figure. 6. Static wall pressure profiles of the spike and the cowl for various mesh resolutions.

3. RESULTS AND DISCUSSION

3.1. Comparison with experiments and validation

Computational results for the three specific NPRs of 2.10, 2.57, and 3.82 have been compared to the data taken from the experiments for the same NPR conditions. The comparisons of flow separation patterns and static wall pressure profiles are shown in Fig. 7. In Fig. 7, Plug-Num and Plug-Exp represent numerical calculation and experiment of plug. The photographs on the second row show the experimental schlieren images, while the ones on the first row show the computational results at the same NPRs. For the three NPRs, three different types of flow separation patterns as well as basic flow characteristics that can be observed are the external jet boundary developing from the cowl lip, over-expansion shocks formed on both the cowl and spike surfaces, and the interaction between them. The computational schlieren clearly replicates these salient flow features. However, an RSS characteristic, separated wake being reattached because of the Coanda effect[41], is also captured in the computation at NPR = 2.10 while an FSS condition was reported by Verma.[4] The numerical schlieren shows the presence of wake reattachment on the spike surface which cannot be distinguished in the experimental schlieren image, and a detailed analysis on the shock physics can be found in a later section. In addition, the experiments used both vertical and horizontal configurations of the knife edge in the schlieren setup. The vertical knife edge helped to identify horizontal gradients, whereas the horizontal knife edge helped to identify the vertical gradients. Note that the upper and lower halves of the experimental flow schlieren exhibit an axisymmetric nature of the shock pattern which indicates that the axisymmetric computational methodology here is advisable to accurately predict the flow physics.

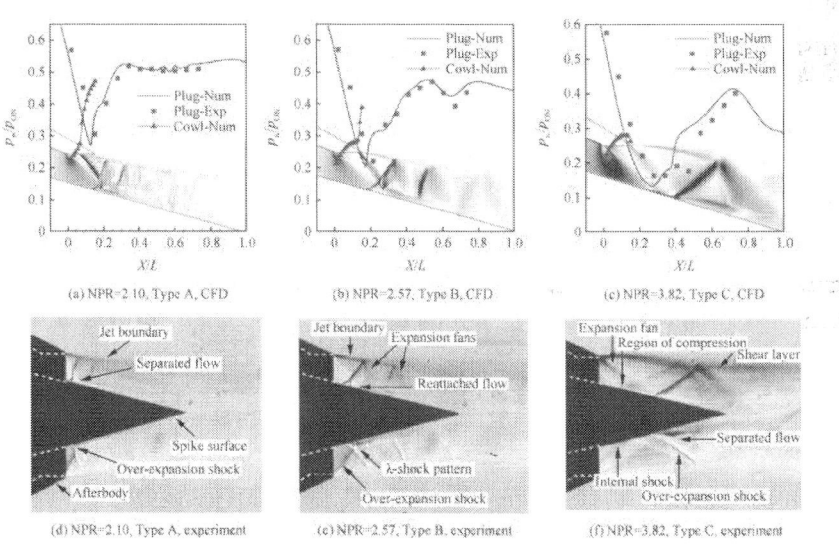

Figure. 7. Comparison of static wall pressure distributions and shock-induced flow separation physics between numerical calculations and wind tunnel experiments with no freestream by Verma,[4] Case 1-1,p_a = 101325 Pa.

In terms of static wall pressure distribution, as shown in numerical schlieren images on the top row, the static wall pressure distributions along the spike surface are compared. All three plots show good agreement with the experimental results in both shock location and pressure distribution. However, the shock location is under-predicted for all the three NPRs and the miss-prediction expands as the NPR changes from low to high values. This discrepancy may contribute to miss-mimicking the experimental condition. In the experiments, the ambient pressure p_a was measured on the afterbody, 15-mm upstream from the cowl lip, by varying the jet stagnation pressure p_{oj} to vary the NPR, and the local ambient pressure p_a decreased with each increase in p_{oj} because of the jet entrainment effect. Thus, the NPR defined as the ratio of the upstream total pressure to the ambient pressure, is therefore different for the experiments and the present numerical simulations. Overall, the computational results agree well with the experimental data in both shock physics and static wall pressure distribution, and the characteristics described by both the simulations and the experiments are indicative of the flow physics that are observed at the three NPRs.

3.2. Flow separation and shock structure progression in the nozzle

To understand the details of the flow separation behavior as well as the shock structure motion in the nozzle, computations when the NPR is increased from low to high values are conducted and the numerical schlieren images of shock structures constructed from computational results are summarized in Fig. 8. In this section, the numerical schlieren pictures are also contoured by the plots of the absolute values of the first gradients of density at the grid nodes, and additionally, static wall pressure distribution plots along the cowl and spike surfaces are superimposed on the schlieren pictures in order to help understand the flow physics.

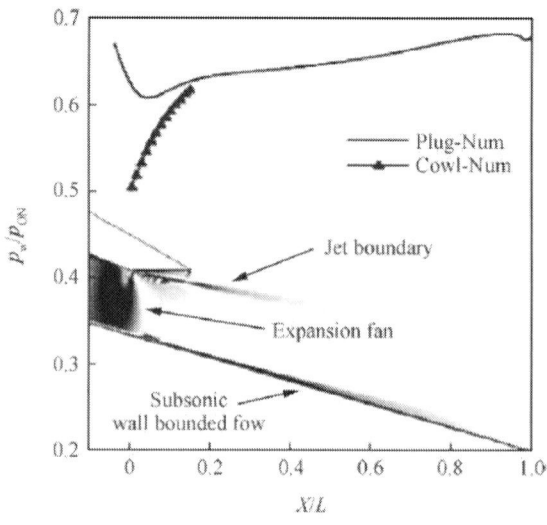

(a) NPR=1.60, no separation

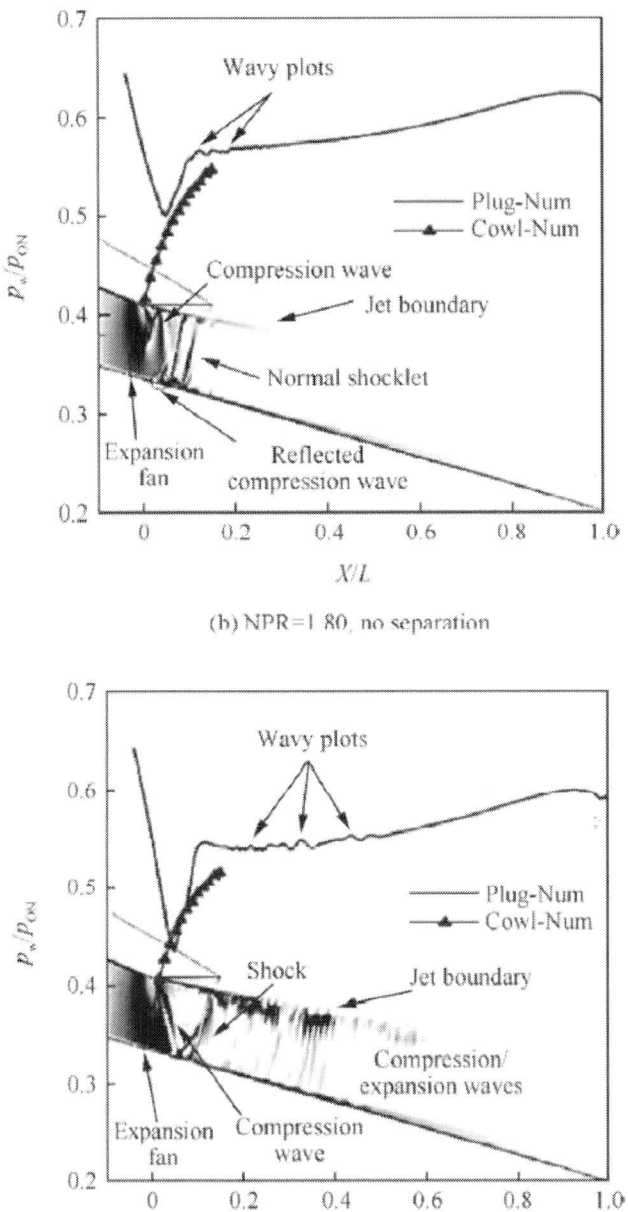

(b) NPR=1.80, no separation

(c) NPR=1.90, no separation

Figure. 8. Shock structure motion as the NPR is increased from 1.60 to 1.90 at the sea-level atmospheric conditions, Case 1-1, $p_a = 101325$ Pa.

We begin the discussion with NPRs ranging from 1.60 to 1.90 when the shock structure resides within the cowl, but was not captured in the experiments.[4] A prominent feature that can be found in the computational schlieren is that a fully separated shear layer emerges from the throat of the cowl-side, and the shear layer as well as the spike surface forms an aerodynamic diverging passage downstream from the throat region. This phenomenon is absent in an over-expanded conventional nor the similar shrouded plug nozzle reported by Kapilavai et al.[7]Fig. 9 gives a line diagram of the nozzle throat design, in which the aerospike nozzle considered presently has a sharp change in the slope at the throat on the cowl-side, and a turning angle up to 20.37° makes the nozzle inlet total pressure here not high enough to expand the jet to form a wall-bounded flow on the cowl surface by Prandtl–Mayer expansion. As the NPR increases, the separated shear layer moves toward the cowl surface and the flow in the above aerodynamic diverging passage exhibits a multitude of shock structures. Detailed flow physics is seen more clearly in Fig. 10.

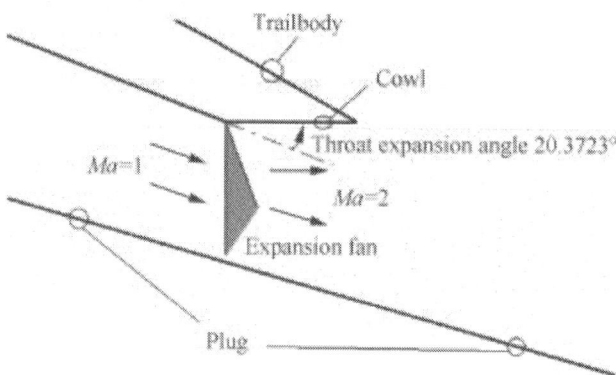

Figure. 9. Line diagram of the annular conical aerospike nozzle throat configuration.

For the NPR = 1.60 shown in Fig. 8 (a), note that the static wall pressure at the throat region is over 0.528 times of the nozzle inlet total pressure, and the flow near the throat region is still unchoked, so no shock structure is captured in the flowfield at this NPR condition. When the flow negotiates the throat region, it goes back to a subsonic wall-bounded flow condition on the spike surface. As the NPR increases to 1.80, the throat region starts to choke as shown in Fig. 8(b) and Fig. 10(a), respectively. The shock exhibits an apparent asymmetric structure like a normal shocklet structure on the spike surface, which is similar to what can be discerned in a transonic diffuser at low NPRs[12]while a serial of expansion and compression waves or shocks appears on the cowl-side flow region. As shown in Fig. 10(a), the flow is compressed immediately by the separated shear layer just after an expansion to Mach number 1.24 through an expansion fan at the sharp corner, and then again expands to an over-expansion shock downstream. The expansion fan that emerges from the over-expansion shock foot over-expands the jet finally to a normal shocklet in the diverging section. As the NPR increases to a slightly higher value of 1.90 as shown in Fig. 8(c) and Fig. 10(b), the normal shocklets coalesce in a single normal shock downstream and occupy the entire cross section. However, the shock is still not strong enough to induce the spike wall boundary separation. In this NPR case, the compression waves induced by the separated shear layer now coalesce in an oblique compression shock which travels across the entire stream without the interception of the over-expansion shock observed in the case of NPR = 1.80. This oblique compression shock is incident on the spike surface and the reflected shock from the spike surface hits the normal shock downstream. Another feature that can be noticed is the presence of weak expansion and compression waves in the aft region of the normal shock, which seem to be a manifestation of the natural convergent–divergent passage formed downstream from the normal shock, and evidence can also be found from the wavy plots of the static wall pressure distribution along the spike surface.

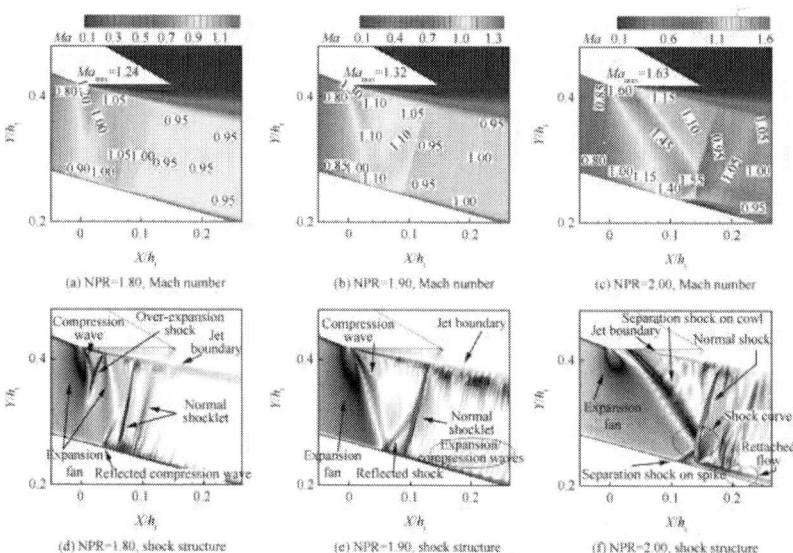

Figure. 10. Close-up view of flow characteristics near the throat region at low NPR conditions for Case 1-1,p_a = 101325 Pa. Pictures are contoured by the data constructed from the computational results with the Mach number for the upper row while the absolute values of the first gradients of density at the grid nodes are on the bottom.

The separated shear layer finally reattaches on the cowl surface when the NPR is about 1.96, and the flow undergoes shock-induced flow separation on both the cowl and spike surfaces when the NPR increases to 2.00, as shown in Fig. 11(a). A close-up view of the computational result can be found in Fig. 10(c). At this point, the shock structure consists of oblique shocks starting from both the spike and cowl surfaces, and these two oblique shocks occupy the downstream normal shock where their interaction makes the normal shock of the spike-side flow region to curve. The flow along the spike surface separates and reattaches in a short distance forming a small recirculation bubble while the flow is fully separated over the entire length of the cowl just aft of the oblique shock.

As the NPR increases from 2.00 to a slightly higher value, the shock structure in the mean flow starts to move out of the cowl which can be captured by the experimental schlieren images. Increasing the NPR from 2.10 to 3.82 changes the angles of the oblique shocks on both the cowl and spike surfaces and their interaction at the internal nozzle operates from the over-expanded to the under-expanded condition. As indicated by Verma[4], this produces different shock structures as well as shock-induced flow separation behaviors on both the cowl and spike surfaces, which can be classified into three types. Firstly, in the low NPR = 2.10 case, the numerical schlieren shows a shock/boundary layer interaction similar to that seen in planar supersonic nozzles.[12 and 43] The shock structure exhibits a Mach reflection,[44] and oblique shocks starting from both the spike and cowl surfaces anchor a normal Mach stem in the mean flow. The flow along the spike surface separates and reattaches again in a short distance forming a small recirculation bubble, and evidence can be found in figures illustrated in a later section, which are contoured by the static wall pressure distribution and axial value of wall shear stress plots along the spike surface, thus exhibiting a restricted shock flow separation, RSS, condition. However, Verma has reported an FSS characteristic for this NPR case in Ref.[4], and this discrepancy may be produced by the undistinguishable wake reattachment in the experimental schlieren image. At this point, the flow separation behavior is not a classical RSS condition, while the mechanism of the wake reattachment is more likely the Coanda effect.[45 and 46] The asymmetric flow passage leads to a radial momentum toward the spike-side flow region and the separated shear layer is very close to the spike surface, so the lower pressure in the separation zone works like an extra suction force, in addition to that of the original Coanda effect on the supersonic jet side, ensuring the separated wake to adhere to the wall.

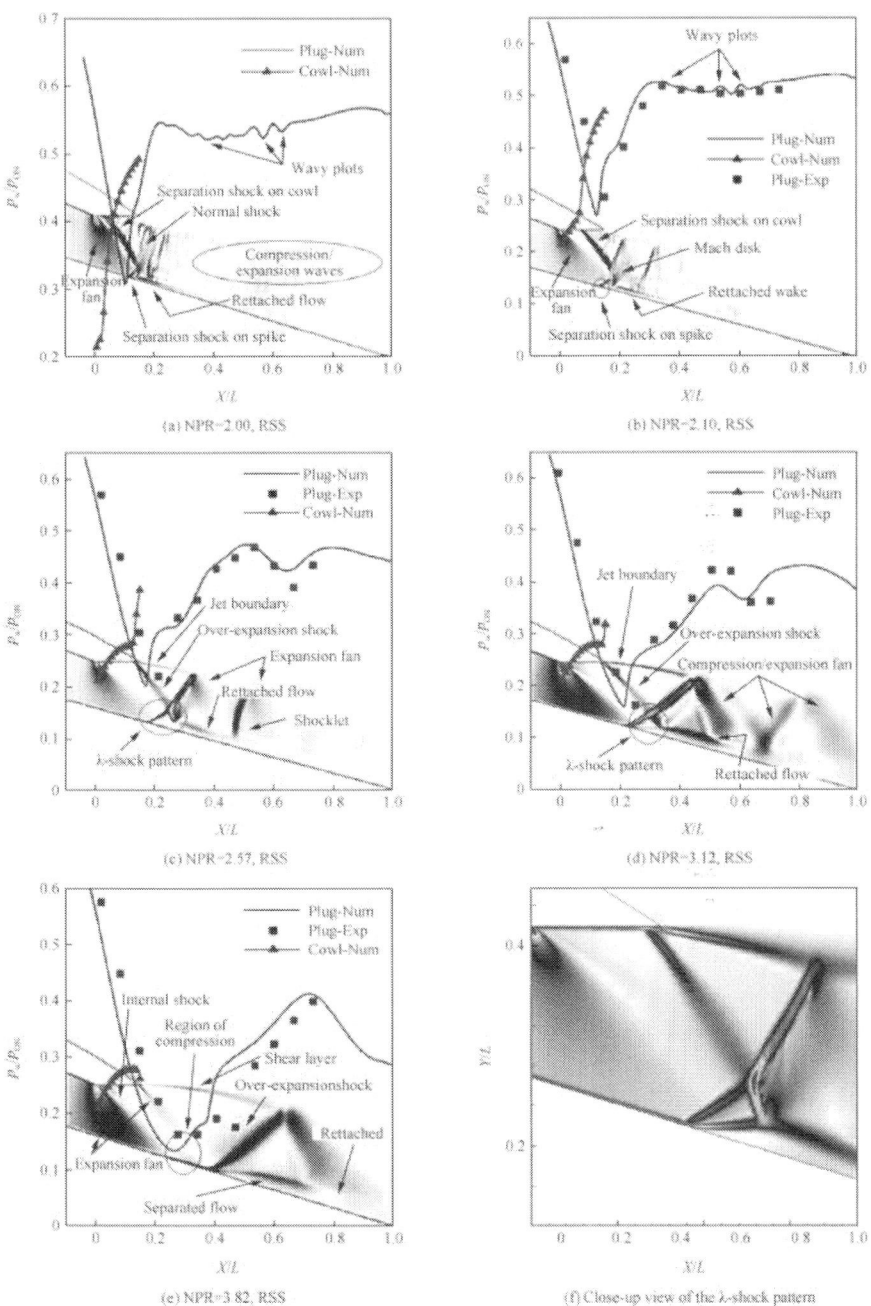

Figure. 11. Shock structure motion as the NPR is increased from 2.00 to 3.82 and a close-up view of the λ-shock pattern for an NPR of 2.57 at the sea-level atmospheric conditions, Case 1-1, $p_a = 101325$ Pa.

At an intermediate value between 2.57 and 3.12, a second shock structure can be discerned. The length of the normal portion of the normal shock decreases with the increase of the NPR, because the strength of the oblique shock on the cowl surface weakens while the interaction between the oblique shock on the spike surface and the normal shock forms a lambda shock pattern on the spike. Detailed flow physics of such aλ-shock pattern for an NPR of 2.57 is seen more clearly in Fig. 11(f). The flow then reattaches which increases the local static wall pressure above ambient, as shown inFig. 11(c)–(d). For NPR = 3.82, a third flow condition is produced when the internal nozzle starts to operate in a little under-expanded condition shown in Fig. 11(e). The expansion fan emerges from the cowl lip and over-expands the flow along the spike surface, resulting in the forming of an oblique over-expansion shock, which hits the free shear layer from the cowl lip and induces an expansion fan at the intersection point. The shock pattern on the spike surface that encloses a recirculation bubble indicates that the nozzle is still in the RSS regime. Another feature that can be discerned from the schlieren is that the internal shock originating from the sharp corner of the throat impinges on the spike surface upstream from the separation shock, forming a region of compression that induces a pressure bump shown in Fig. 11(e).

At higher NPRs, the expansion fan impinges further downstream causing the over-expansion shock as well as the flow reattachment point to move downstream. At NPR = 5.22, the over-expansion shock starts to induce a free shock separation condition on the spike, as shown in the axial wall shear stress plots in later section. The length of the separation bubble just covers the entire portion of the spike downstream from the incident separation point; detailed analysis on this flow condition is followed in a later section.

For NPRs above 5.22 through 8.40, the free shock separation is the mode of separation on the spike surface. Increasing the NPR to 5.75 continues to weaken the separation shock on the spike while strengthening the internal shock as well as the expansion fan from the cowl lip, and the region of compression induced by the impingement of the internal shock now produces a distinct pressure bump, as shown in Fig. 12(a). Additionally, an important phenomenon in these NPR cases that should not be neglected is the motion of the separated shear layer away from the spike surface. This behavior indicates a higher pressure in the separation zone to push the shear layer towards the mean flow. The mechanism of the phenomenon may be interpreted by the impinging jet effect. The spike considered here has a simple conical configuration where the annular jet that expands from the end of the spike has an impinging angle of 30°. Even at present the NPR is slightly below the design point, and the separation bubble is enclosed by the annular main flow producing a higher local pressure up to 1.7 times of the ambient pressure as shown in Fig. 13(e). However, this impinging region will not be present if the spike has been contoured.

At NPR's above 7.75, a new type of shock structure is produced when the internal nozzle starts to operate in a highly under-expanded condition in which an interception shock is generated at the cowl lip while a reflected shock emerges from the spike surface where the internal shock originating from the sharp corner of the throat impinges, as shown inFig. 12(b). Increasing the NPR to the design point continues to strengthen these shocks. A stronger expansion fan coming from the cowl lip over-expands the flow along the spike surface and leads to the free shock separation formation even in the case of the designed condition shown in Fig. 12(c). For NPR's above 9.87 shown in Fig. 12(d), the pressure on the spike surface is high enough and no flow separation can be captured.

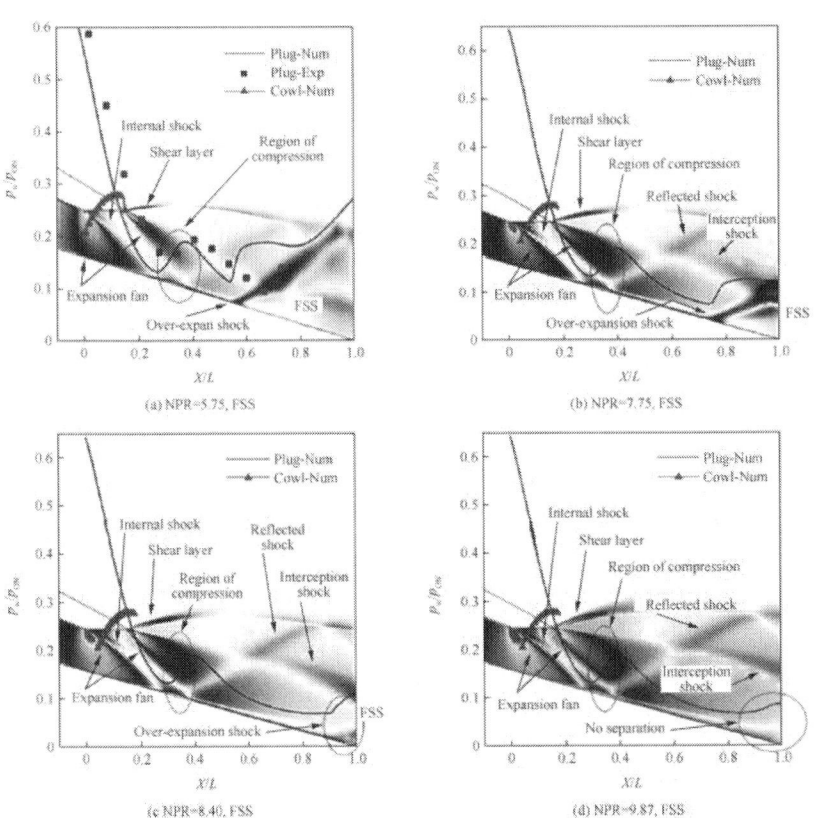

Figure. 12. Shock structure motion as the NPR is increased from 5.75 to 9.87 at the sea-level atmospheric conditions, Case 1-1, $p_a = 101325$ Pa.

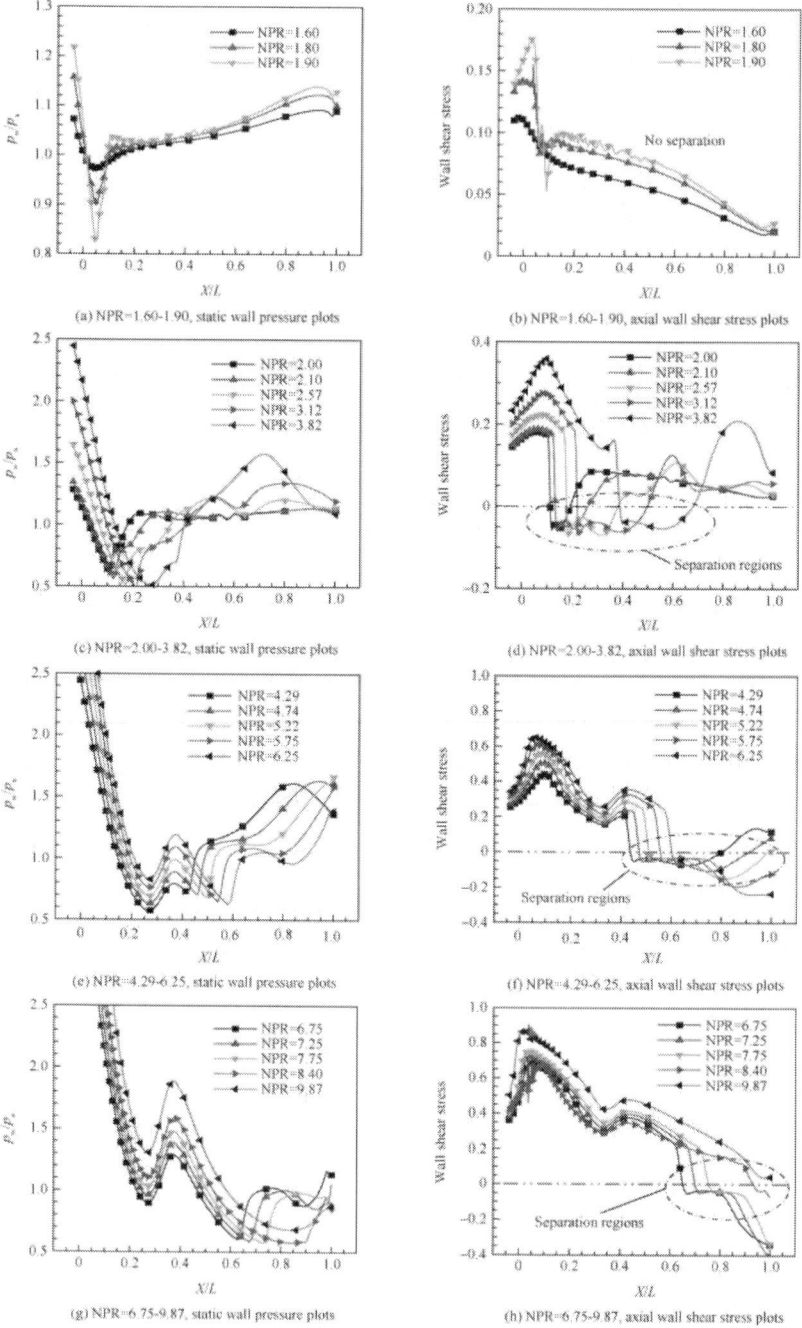

Figure. 13. Streamwise distribution of the static wall pressure and the axial wall shear stress along the spike surface for different NPRs in the sea-level atmospheric conditions, Case 1-1, $p_a = 101325$ Pa.

It is important to note here that unlike a contoured spike configuration in a conventional aerospike nozzle, in which the nozzle achieves expansion to the ambient condition by means of a centered expansion fan generated at the cowl lip and terminated at the spike end at the design condition, the expansion fan hits at shorter distances on the spike, and as a consequence, a series of expansion/compression waves continues till the spike end. In the present case, the non-uniform flow exhausting from the internal nozzle in addition to the conical configuration of the spike surface makes the spike geometry incapable of canceling out all the impinging waves. Firstly, the flow at the exit of the internal nozzle is not uniform, due to the asymmetric expansion fans coming from the nozzle throat and the following compression shock generated by the sharp change in slope at the throat section. Secondly, when the internal nozzle starts to operate in under-expanded conditions, additional asymmetric expansion fans and interception shock are generated from the cowl lip. Finally, the asymmetric flow passage in addition to that impinging jet effect induces a higher back pressure. The result of these phenomena is the formation of stronger compression shock and sustaining over-expansion of the flow along the spike surface to a shock-induced flow separation even at the design condition.

3.3. Static wall pressure and axial wall shear stress profiles

The numerical schlieren images of shock structures constructed from computational results obtain additional shock physics and associated flow separation behavior at various NPR's against the experimental methodology. In particular, the presence of a restricted shock separation regime at low and moderate NPRs while a free shock separation regime at high NPR conditions is recognized. In order to further confirm these flow characteristics, the static wall pressure as well as the axial wall shear stress distribution on the spike is instructive. The plot of the axial wall shear stress on the spike can clearly show the separation as well as the reattachment point induced by the incident and separation shocks by that the value decreasing to below zero indicates that the boundary separates at the present point while increasing to above zero indicates that the boundary reattaches again. Fig. 13 shows the plots of the predicted static wall pressure and the axial wall shear stress distribution on the spike. As the nozzle exhibits a multitude of shock structures and associated flow separation regimes at various NPR's, the figures show a progression of NPRs for the entire range. The throat is located at $X/L = 0$, while the static wall pressure, the axial wall shear stress, and the X co-ordinate have been normalized by the ambient pressure, the maximum value of the axial wall shear stress, and the spike length, respectively.

The plots in Fig. 13(a) and (b) correspond to that of no flow separation regime for NPR's below 1.96, and the flow expands from a subsonic condition at an NPR of 1.60 to a normal shock condition at an NPR of 1.90. Looking at the plot with an NPR of 1.90 for the spike, we notice a bump in the static wall pressure at $X/L = 0.09$, for which as in our earlier discussion, this pressure increase is induced by the normal shock which is not strong enough to induce the flow boundary separation, and it is also evident in the axial wall shear stress distribution that the value stays above and beyond zero along the entire spike as shown in Fig. 13(b). Another feature that can be discerned in the pressure distribution is the pressure adaptation to the ambient condition, in which the spike is enclosed by the annular jet and induces a back pressure higher than that in the ambient condition, and then the static wall pressure increases above and beyond the ambient pressure just aft in the throat region monotonically. This scenario is absent in a conventional bell-type nozzle during transonic conditions.

Fig. 13(c) shows the static wall pressure distribution in the restricted shock separation regime for NPR = 2.00, 2.10, 2.57, 3.12, 3.82. It is evident from the axial wall shear stress plots that the nozzle really exhibits an RSS regime at these NPRs, and in fact the details of the flow pattern cannot be recognized by the experimental methodology.[4] More than five types of shock structures are captured at these NPRs, however, only the moderate NPR = 2.57, 3.12, exhibit a conventional RSS regime while the lower, NPR = 2.00, 2.10, and higher, NPR = 3.82, NPRs are not classical RSS conditions as seen earlier. In case of the conventional RSS regime, the shock exhibits a λ structure and the pressure rises after the first leg of the λ-shock and then forms a plateau which ends when the reflected shock, the second leg of the λ-shock, hits the recirculation bubble. Immediately after this, the pressure increases till the reattachment point and then monotonically decreases as the flow expands downstream from reattachment. However, using the present case of NPR = 3.82 as an example, the expansion fan emerging from the cowl lip over-expands the flow along the spike surface and results in forming of a single oblique over-expansion shock; the pressure rises after the separation shock ($X/L = 0.4$) and continues to increase till the reattachment point ($X/L = 0.66$), and then decreases downstream from reattachment, while no plateau can be discerned along the recirculation bubble.

Fig. 13(e) shows the static wall pressure distribution progression during the flow separation pattern transition regime for NPRs of 4.29–6.25 when the shock structure transforms from the restricted shock separation regime to the free shock separation regime. The transition happens at an NPR of 5.22 as seen earlier, and is also evident in the axial wall shear stress plots shown in Fig. 13(f). In contrast to flow separation transition phenomena in a bell-type nozzle, there is no distinct step change in the evolution of the static wall pressure on the spike. This difference is not entirely clear but is consistent with our computation upon the grid density effect study later. Examination of the static wall pressure plots shows a region of compression (seen as a hump) starting from X/L of 0.26–0.27. The hump grows up with increasing the nozzle pressure ratio while its starting point has slight excursion. In addition, the static wall pressure profile shows similar evolution as the one for a classical RSS condition that the pressure rises after the separation shock and then forms a plateau which ends in a distance, and immediately after this, the pressure increases till p_w/p_a is over 1.5. Moreover, this scenario indicates a different free shock separation behavior from the one in a conventional axisymmetric bell-type nozzle.

In the final figures of the series, Fig. 13(g) and (h), these NPR's shown exhibit that the flow condition on the spike transforms from the free shock separation at NPRs of 1.60–8.40 to full expansion at an NPR of 9.87 although the static wall pressure at the end of the spike surface is still below that in the ambient condition. The plots of the axial wall shear stress show clearly the progression of the separation point. The region of compression on the spike now shows a distinct static wall pressure plot hump. By using both the static wall pressure non-dimensionalized with respect to the ambient condition and the axial wall shear stress profile, the fact that the expansion fan coming from the cowl lip over-expands the flow along the spike surface and leads to the free shock separation formation even in the case of the designed condition can be clearly recognized.

3.4. Prediction of the flow separation behavior

As flow separation may lead to performance losses and undesired high nozzle structural loads,[47, 48 and 49] an accurate separation criterion is crucial. A series of separation criterion has been developed with increasing knowledge and availability of experimental data. However, most of the historical data are predominantly for conical or bell-type nozzles with an axisymmetric configuration, and thereby applicability on the present asymmetric nozzle needs to be validated. Fig. 14 plots the separation data (p_{sep}/p_a or p_{sep}/p_{ON}) as a function of the corresponding wall separation Mach number, Ma_{sep}, or the NPR, on the spike surface as well as the cowl wall, where the wall separation Mach number is based on the isentropic ratio of p_{ON}/p_{sep}. Three separation criteria are plotted for comparison: the well-known Schmucker criterion[50] (1), the separation criterion (2) for turbulent nozzle flows suggested by Stark et al. [51 and 52], and the separation criterion (3) suggested by Ge et al.[53] recently based on flow separation data in asymmetric ramp nozzles.

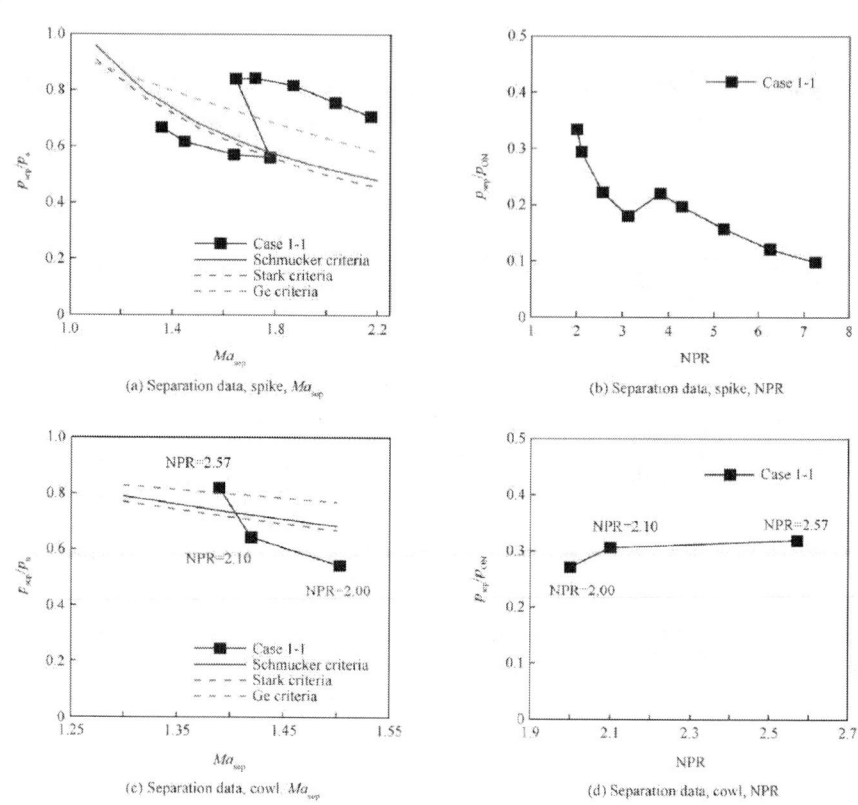

(a) Separation data, spike, Ma_{sep}

(b) Separation data, spike, NPR

(c) Separation data, cowl, Ma_{sep}

(d) Separation data, cowl, NPR

Figure. 14. Comparison of separation data plots on the spike and cowl surfaces at the sea-level atmospheric conditions, Case 1-1, $p_a = 101325$ Pa as a function of Ma_{sep} and NPR, respectively.

Equation (1)

$$\frac{p_{sep}}{p_a} = \left(1.88 Ma_{sep} - 1\right)^{-0.64}$$

Equation (2)

$$\frac{p_{sep}}{p_a} = \frac{1}{Ma_{sep}}$$

Equation (3)

$$\frac{p_{sep}}{p_a} = -1.76\left(0.47Ma_{sep}^{0.45} - 1\right)$$

Looking at the plots of separation data on the spike wall shown in Fig. 14(a) and (b), the data affected by the region of compression, the over-expansion by the expansion fan generated at the cowl lip, and the enclosed back pressure environment by the annular jet for NPR $\geqslant$ 3.82 when the internal nozzle starts to operate in an under-expanded condition are clearly pointed out in the plots. A region covers nozzle pressure ratios of 3.12–3.82, corresponding to the wall separation Mach number of 1.65–1.80, and divides the data-set into two regions. The data show a strong variation and are bounded below by all the three separation criteria for NPR $\leqslant$ 3.12 and above by the criteria for NPR $\geqslant$ 3.82. At first sight, an important separation behavior transformation seems to happen in between as discussed earlier. The data of Schmucker and Stark criteria show below NPR $\leqslant$ 3.12 a trend which seems to be a sight parallel shift and reproduces the separation pressure with a reasonable accuracy for this NPR regime, while failing at the prediction for the NPR $\geqslant$ 3.82. The separation criterion (3) suggested by Ge et al. fails to reproduce the separation pressure data for both of the two regions. This is a significant hint that the asymmetric nozzle configuration may affect the flow separation behavior in the present aerospike nozzle and induce the miss-prediction by the separation criteria developed from separation data in conical or bell-type rocket nozzles.

Some interesting effects are also pointed out in the plots of separation data on the cowl wall shown in Fig. 14(c) and (d). For NPR's between 2.00 and 2.57, the flow separation on the cowl wall is always in the FSS regime as seen earlier. However, with the interception effect of the oblique compression shock induced by the separated shear layer at the sharp corner, the flow separation on the cowl surface exhibits abnormal free shock separation behavior. The separation pressure, p_{sep}, show an opposite trend as a function of NPR that increasing the NPR strengthens the upstream oblique compression shock, leading to a higher local static wall pressure and total pressure loss, and then to a lower wall separation Mach number, Ma_{sep}. Again, all the three separation criteria fail to reproduce the separation data on the cowl wall.

3.5. Gas density effect

The above discussion gives a detailed view of the shock physics and flow separation behavior as the NPR is increased from low to high values for the sea-level condition. Now, we compare the gas density effect on the above-mentioned flow characteristics at specific NPRs for the three ambient pressure conditions. Recent studies54 [and] 55 on a thrust-optimized parabolic nozzle have reported significant variations in results when comparing data from tests conducted in sea-level atmosphere with those inside a high-altitude chamber. The series of pictures illustrated in Fig. 15 shows the predicted shock patterns at the three different ambient pressure conditions for NPR = 2.00, 2.10, and 2.57, when more complex shock/boundary layer interaction is expected for these highly over-expanded conditions. It is noteworthy here that the results of shock-induced flow separation structures and location as well as the static wall pressure profiles variations with the three different ambient pressure conditions show a similar tendency as those for NPR = 2.57. As a consequence, the results for higher NPRs are not included in the paper. Pictures are also contoured with the same levels by the data constructed from the computational results with the absolute values of the first gradients of density at the grid nodes. At first sight, the shock structure is somewhat less distinct in the high-altitude simulations, indicating a weaker shock system for lower density flows.

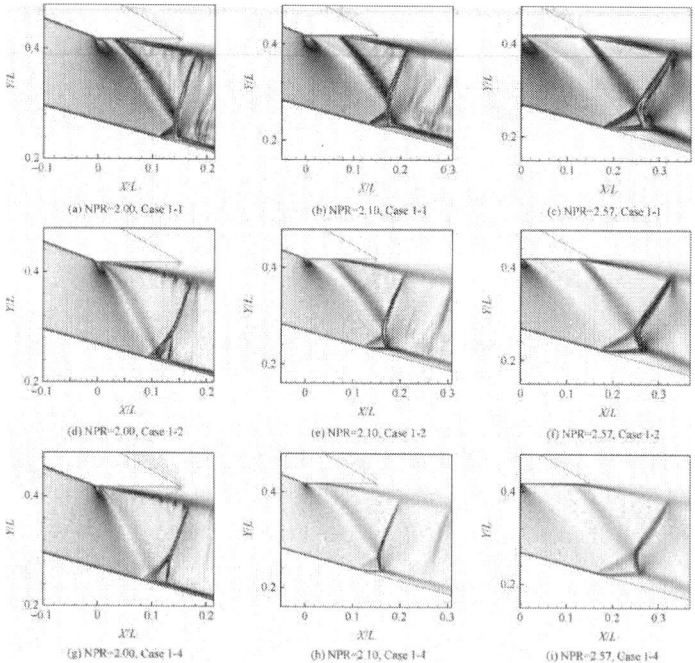

Figure. 15. Comparison of the predicted shock structures at different ambient pressure conditions for NPR = 2.00, 2.10, and 2.57. Pictures are contoured at the same levels by the data constructed from the computational results with the absolute values of the first gradients of density at the grid nodes.

Looking at the pictures on the left of Fig. 15(a), (d), and (g) from top to bottom for an NPR of 2.00, we note that the predicted shock patterns and associated flow separation on both the spike and cowl surfaces in the two high-altitude simulations exhibit distinct differences in contrast to the sea-level condition, suggesting a strong gas density effect at this NPR condition. As discussed earlier, this particular NPR of 2.00 lies in the shock transition from a normal shock to Mach reflection on the spike surface while the separation point jumps from the throat's sharp corner to the downstream cowl extension. For flows in the high-altitude simulations, the shock structure generates a λ pattern enclosing a smaller recirculation zone at the spike. The oblique compression shock starting from the throat's sharp corner is weak now and hits on the first foot of the lambda shock; the separation shock is modified in the vicinity of the wall by the intersection between these shocks.

In terms of cowl side, decreasing the ambient pressure pushes the separation point on the cowl surface back to the throat's sharp corner, and it freezes exactly when the ambient pressure increases from 25% to 50% at high-altitude simulation conditions, suggesting that a higher inlet total pressure is needed to push the separation point to jump downstream into the cowl extension for lower density flow. This phenomenon is similar to the gas density effect on the sneak transition process in dual-bell nozzles, and recent studies[54] on a subscale dual-bell nozzle have reported that the separation point for tests inside a high-altitude simulation chamber moves into the region of wall inflection much earlier and stays there for a much longer time. This delays the process of transition and hence increases the NPR of dual-bell transition as p_{ON} is decreased for tests inside the high-altitude simulation chamber. For results at higher nozzle pressure ratios of 2.10 and 2.57 shown in the middle and right list of Fig. 15, the comparison between the predicted shock patterns at the three ambient conditions shows a close one at these higher NPRs. One difference between the three results is that the separation point shows a distinct excursion upstream on both the spike and cowl surfaces as the ambient pressure decreases. The other feature that can be discerned is a smaller excursion of the separation point at a higher NPR of 2.57 than that observed for an NPR of 2.10. This suggests that the gas density effect may be weakening at a higher Mach number flow regime, when the flow Reynolds number increases with the increase of nozzle pressure ratio as well as flow Mach number, and the Reynolds number has been reported to have a significant influence on the shock/boundary layer interaction.[54]

Fig. 16(a)–(c) plot the streamwise distribution of the non-dimensional mean static wall pressure for the three NPRs at different ambient conditions. Again, the plot of the axial wall shear stress along the spike is used here to interpret the excursion behavior of the separation shock as shown in Fig. 16(d)–(f). It may be noted that approximately similar values of the static wall pressure are being experienced irrespective of the simulated ambient pressure conditions except in two regions. One difference is that the lowest pressure value just before the separation shows an increase with a decrease in the ambient pressure. Therefore, a lower pressure rise is needed for the static wall pressure adaptation to the ambient condition, resulting in a weaker separation shock for the lower-density flow, as shown in the predicted shock patterns. The second difference is the absent wavy plots of the static wall pressure distribution for the two high-altitude simulation cases. This suggests that, for lower gas density, the weak expansion and compression waves in the aft region of the main normal shock are also weakened by the separation shock and even disappear for extremely low ambient pressure conditions. A distinct decrease of the axial wall shear stress with a decrease in the ambient pressure is shown in the axial wall shear stress plots, Fig. 16(d)–(f). The excursion upstream from the separation point location, the reduction in the lateral extent of the separation zone, and the weaker influence on the above two features with a decrease in the ambient pressure are shown clearly in the pictures.

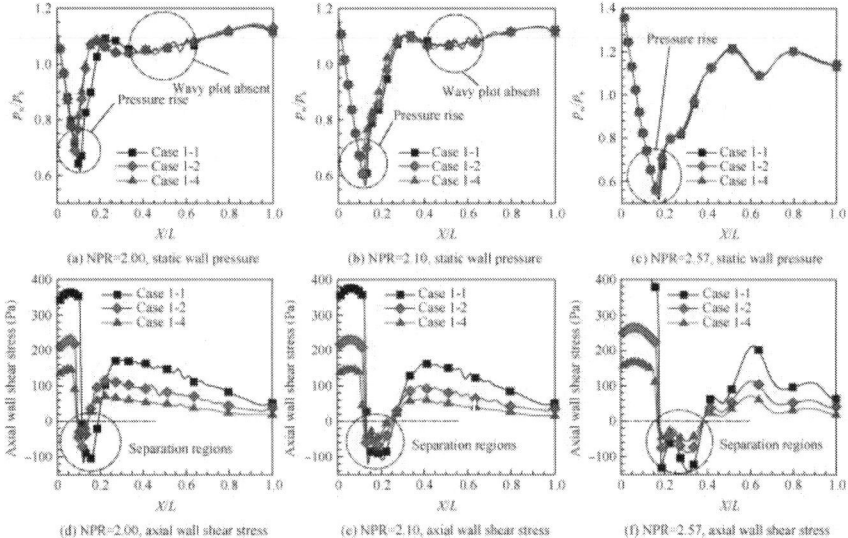

Figure. 16. Comparison of the static wall pressure and the axial wall shear stress distribution along the spike wall at different ambient pressure conditions for NPR = 2.00, 2.10, and 2.57.

Fig. 17 shows the variation of the streamwise distribution of the non-dimensional mean static wall pressure and the axial wall shear stress along the cowl surface. Because the length of the cowl, l, is smaller compared with that of the spike, the gas density effect is more distinct for the separation behavior on the cowl surface. It can be noted that the static wall pressure distribution shows a larger discrepancy in the separation region, while with a decrease in the ambient pressure, the rate of pressure rise across the separation shock decreases gradually. This feature is less distinct with the increase of the NPR shown in Fig. 17(c). The phenomenon that the location of the incipient separation point excurses back to the throat's sharp corner and freezes exactly even with large changes in the ambient pressure (transition from the 25% to the 50% ambient pressure case) is shown clearly in the axial wall shear stress plots, Fig. 17(d)–(f).

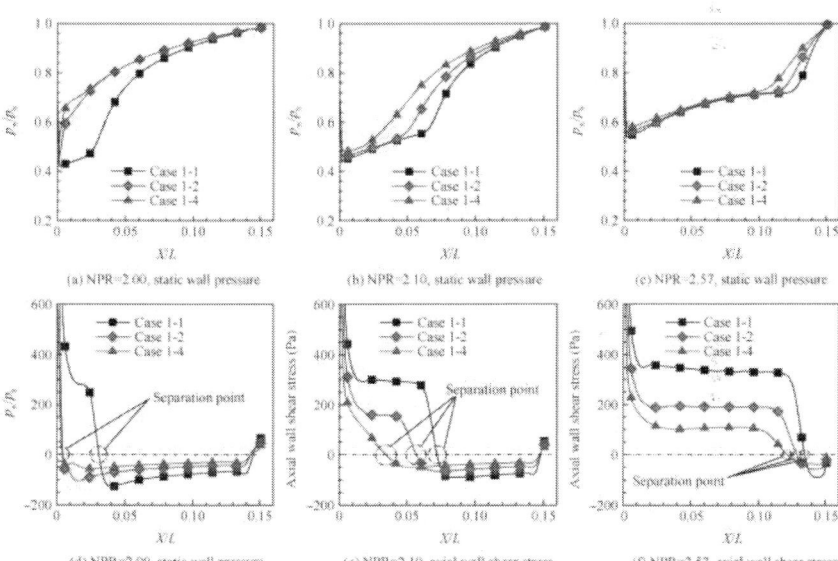

Figure. 17. Comparison of the static wall pressure and the axial wall shear stress distribution along the cowl wall at different ambient pressure conditions for NPR = 2.00, 2.10, and 2.57.

To investigate the reason behind the observed discrepancies, Fig. 18 shows a comparison of the static wall pressure gradient near the separation shock region for the three ambient conditions at NPR = 2.10, 2.57. Here the static wall pressure gradient, dp/dx, is non-dimensional (the static wall pressure, p_w, is non-dimensionalized with respect to the ambient pressure, p_b) and the X co-ordinate is normalized by the annular gap at the throat section h_t. The separation point locates between X/h_t of 0.5–1.0 and 1.0–1.5 for NPR = 2.10, 2.57, respectively. It can be seen that approximately similar values of dp/dx are being experienced outside the separation region irrespective of the simulated gas density. However, as the ambient pressure increases, the value of dp/dx shows a decrease with a decrease in gas density across the separation shock. Studies on bell-type nozzles have also indicated that a distinct excursion of the incipient separation point occurs when the adverse pressure gradient across the separation shock is smaller. [56] This suggests that dp/dx is larger for higher gas density and vice versa, indicating a stronger shock system for higher density flow.

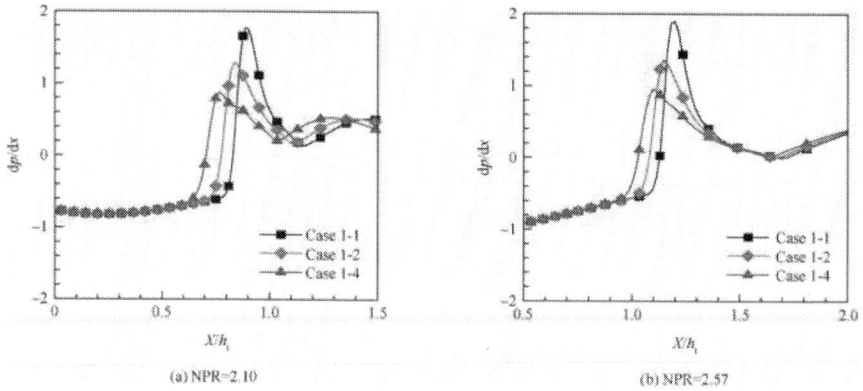

Figure. 18. Comparison of the static wall pressure gradient across separation at different ambient pressure conditions for NPR = 2.10, 2.57.

During a low-altitude mode with NPR = 2.00, the shock patterns and associated flow separation behavior strongly signify the compression shock and the separated shear layer starting from the throat's sharp corner. As discussed earlier, once the throat region has reached the supersonic regime, the wall inflection at the sharp corner that controls separation occurs over a wide range of NPRs because a large pressure gradient prevalent at the corner allows only a small movement of the separation point with large changes in the NPR. In particular, the gas density effect delays this process because of lower dp/dx. Additional calculations and experimental tests on high-altitude simulation are needed to get deeper insight into the aerodynamic mechanism behind the observed gas density effect.

4. SUMMARY AND CONCLUSIONS

1. The computational results agree well with the experimental data in both shock physics and static wall pressure distribution, indicating that the axisymmetric computational methodology here is advisable to accurately predict the flow physics. The progressively increased excursion in the plots of static wall pressure distribution with an increase in the NPR may contribute to slightly miss-mimicking the experimental condition because of the jet entrainment effect.

2. The annular conical aerospike nozzle is observed to be dominated by shock/shock and shock/boundary layer interactions at all calculated NPRs, and the shock physics and associated flow separation behavior are quite complex. Increasing the NPR changes the operating condition of the internal nozzle as well as the basic flow physics such as the expansion fans, the separated shear layer, and the over-expansion shock on both the spike and cowl surfaces. As the internal nozzle operates from the highly over-expanded to the over-expanded condition and then to the under-expanded condition, the flow condition on the spike exhibits a normal shock structure with no flow separation for $NPR < 2.0$, a multitude of shock structure transitions with the restricted shock separation for $2.0 \leqslant NPR \leqslant 5.22$, and an oblique over-expansion shock with the free shock separation for $5.22 < NPR \leqslant 8.40$ regimes. This shock and separation transitions are absent in optimized bell-type nozzles nor the reported shrouded plug nozzle, and the identification of these three regimes helps in further investigation on the unsteady fluid dynamics and related side loads generation. The computational results provide additional flow characteristics against the experimental data; in particular, the shock/shock and shock/boundary layer interactions with the restricted shock separation at highly over-expanded conditions and the free shock separation behavior at higher NPRs than those at design conditions are excluded in the experimental study.

3. The separation data show that a strong variation for the spike, a region that covers NPR of 3.12–3.82, corresponding to the wall separation Mach numbers of 1.65–1.80, divides the data-set into two regions. All the three separation criterias fail to reproduce the separation data of the spike for the $NPR \geqslant 3.82$, when the internal nozzle starts to operate in under-expanded conditions. On the other hand, the flow separation structure on the cowl surface is always in the FSS regime. However, with the interception effect of the oblique compression shock induced by the separated shear layer at the sharp corner, the separation pressure, p_{sep}, shows an opposite trend as a

function of the NPR in contrast to that in over-expanded conventional bell-type rocket nozzles. We conclude that the separation criteria developed from separation data in conical or bell-type rocket nozzles may be inapplicable for the prediction of flow separation behavior in the present asymmetric supersonic nozzle.

4. A strong gas density effect has been found at the highly over-expanded condition with NPR = 2.00, when the wall inflection at the sharp corner that controls separation occurs over a wide range of NPRs during this flow regime. However, for results at higher NPRs, the comparison between the predicted shock patterns at the three ambient conditions shows a close one for these higher NPRs regime. This suggests that the gas density effect is going to weaken at a higher Mach number flow regime when the flow Reynolds number increases with an increase of the NPR as well as the flow Mach number. The adverse pressure gradient across the separation shock, dp/dx, is larger for higher gas density and vice versa, indicating a stronger shock system for higher density flow, which may contribute to the observed gas density effect. These results emphasize that the flow separation behavior tests in such an aerospike nozzle inside a high-altitude test facility should be carefully interpreted; in particular, Reynolds number effects are needed to be concerned when comparing results from the sea-level conditions.

REFERENCES

1. Fick M, Schmucker RH. Performance aspects of plug cluster nozzles. J Spacecraft Rockets 1996;33(4):507–12.

2. Rommel T, Hagemann G, Schley CA, Krulle G, Manski D. Plug nozzle flowfield analysis. J Propul Power 1997;13(5):629–34.

3. Hagemann G, Immich H, Terhadt M. Flow phenomenon in advanced rocket nozzles-the plug nozzle. Reston: AIAA; 1998. Report No.: AIAA-1998-3522.

4. Verma SB. Performance characteristics of an annular conical aerospike nozzle with freestream effect. Reston: AIAA; 2008. Report No.: AIAA-2008-5290.

5. Schwane R, Hagemann G, Reijasse P. Plug nozzles-assessment of prediction methods for flow features and engine performance. Reston: AIAA; 2002. Report No.: AIAA-2002-0585.

6. Onofri M. Pug nozzles: summary of flow features and engine performance. Reston: AIAA; 2002. Report No.: AIAA-2002-0584.

7. Kapilavai DSK, Tapee J, Sullivan J, Merkle CL, Wayman TR, Conners TR. Experimental testing and numerical simulations of shrouded plug-nozzle flowfields. J Propul Power 2012;28(3): 530–44.

8. Arens M, Spiegler E. Shock-induced boundary layer separation in overexpanded conical exhaust nozzles. AIAA J 1963;1(3):578–81.

9. Schmucker RH. Flow processes in overexpanded chemical rocket nozzles. Part 2: side loads due to asymmetric separation. Washington, D.C.: NASA; 1984. Report No.: NASA-TM-77395.

10. Nave LH, Coffey GA. Sea level side loads in high-area-ratio rocket engines. Reston: AIAA; 1973. Report No.: AIAA-1973-1284.

11. Nguyen AT, Deniau H, Girard S, de Roquefort TA. Unsteadiness of flow separation and end-effects regime in a thrust optimized contour rocket nozzle. Flow Turbul Combust 2003;71(1–4):161–81.

12. Hagemann G, Frey M, Koschel W. Appearance of restricted shock separation in rocket nozzles. J Propul Power 2002;18(3):577–84.

13. Ostlund J. Flow processes in rocket engine nozzles with focus on flow-separation and side-loads. Mekanik (Stockholm): Royal Institute of Technology; 2002.

14. Chen CL, Chakravarthy SR, Hung CM. Numerical investigation of separated nozzle flows. AIAA J 1994;32(9):1836–43.

15. Gross A, Weiland C. Numerical simulation of separated cold gas nozzle flows. J Propul Power 2004;20(3):509–19.

16. Deck S, Nguyen AT. Unsteady side loads in a thrust-optimized contour nozzle at hysteresis regime. AIAA J 2004;42(9):1878–88.

17. Nasuti F, Onofri M. Viscous and inviscid vortex generation during start-up of rocket nozzles. AIAA J 1998;36(5):809–15.

18. Mori´nigo JA, Salva´ JJ. Three-dimensional simulation of the selfoscillating flow and side-loads in an over-expanded subscale rocket nozzle. Proc IMechE Part G J Aerosp Eng 2006;220(5):507–23.

19. Baars WJ, Tinney CE. Transient wall pressures in an overexpanded and large area ratio nozzle. Exp Fluids 2013;54(2):1–17.

20. Baars WJ, Tinney CE, Ruf JH, Brown AM, McDaniels DM. Wall pressure unsteadiness and side loads in overexpanded rocket nozzles. AIAA J 2012;50(1):61–73.

21. Ostlund J, Damgaard T, Frey M. Side-loads phenomena in highly over-expanded rocket nozzles. Reston: AIAA; 2001. Report No.: AIAA-2001-3684.

22. Stark R, Kwan W, Quessard F, Hagemann G, Terhardt M. Rocket nozzle cold gas test campaigns for plume investigations. Proceeding of the 4th European symposium on aerothermodynamics for space vehicles; 2001 Oct 15-18; Capua, Italy. ESA SP-487, 2002

23. Tomita T, Sakamoto H, Onodera T, Sasaki M, Takahashi M, Watanabe Y. Experimental evaluation of side-loads characteristics on TP, CTP and TO nozzles. Reston: AIAA; 2004. Report No.:AIAA-2004-3678.

24. Mouronval AS, Hadjadj A. Numerical study of the starting process in a supersonic nozzle. J Propul Power 2005;21(2):374–8.

25. Roquefort de TA. Unsteadiness and side-loads in over-expanded nozzles. Proceeding of the 4th European symposium on aerothermodynamics for space vehicles; 2001 Oct 15–18; Capua, Italy. ESA SP-487, 2002.

26. Ostlund J, Muhammad-Klingmann B. Supersonic flow separation with application to rocket engine nozzles. Appl Mech Rev 2005;58(3):143.

27. Page RH, Meyer AP. Hydraulic analog investigation of a plug nozzle. ARS J 1961;31(3):447–8. 28. Angelino G. Approximate method for plug nozzle design. AIAA J 1964;2(10):1834–5.

28. Giel Jr TV, Mueller TJ. Mach disk in truncated plug nozzle flow. AIAA J 1976;13(4):203–7.

29. Berman K, Crimp Jr FW. Performance of plug-type rocket exhaust nozzles. ARS J 1961;31(3):18–23.

30. Rao GVR. Recent developments in rocket nozzle configurations. ARS J 1961;31(3):1488–94.

31. Verma SB, Viji M. Freestream effects on base pressure development of an annular plug nozzle. Shock Waves 2011;21(2):163–71.

32. Verma SB, Viji M. Linear-plug flowfield and base pressure development in freestream flow. J Propul Power 2011;27(6): 1247–58. Numerical investigation of flow separation behavior in an over-expanded annular conical aerospike nozzle 1001

33. Ruf JH, McConnaughey PK. The plume physics behind aerospike nozzle altitude compensation and slipstream effect. Reston: AIAA; 1997. Report No.:AIAA-1997-3217.

34. Menter FR, Kuntz M, Langtry R. Ten years of industrial experience with the SST turbulence model. In: Harjalic K, Nagano Y, Tummers M, editors. Turbunlence, heat, and mass transfer 4. New York: Begell House, Inc.; 2003. p. 625–32.

35. Hunter CA. Experimental, theoretical and computational investigation of separated nozzle flows. Reston: AIAA; 1998. Report No.:AIAA-1998-3107.

36. Xiao Q, Tsai HM, Papamoschou D. Numerical investigation of supersonic nozzle flow separation. AIAA J 2007;45(3):532–41.

37. Xiao Q, Tsai HM, Liu F. Computation of turbulent separated nozzle flow by a lag model. J Propul Power 2005;21(2):368–71.

38. Xiao Q, Tsai HM, Papamoschou D, Johnson A. Experimental and numerical study of jet mixing from a shock containing nozzle. J Propul Power 2009;25(3):688–96.

39. Carlson JR. A nozzle internal performance prediction method. Hampton (Virginia): Langley Research Center; 1992. Report No.: NASA-TM-3221.

40. Wang TS. Transient three-dimensional startup side load analysis of a regeneratively cooled nozzle. Reston: AIAA; 2008. Report No.:AIAA-2008-4300.

41. Sajben M, Kroutil JC. Effect of initial boundary layer thickness on transonic diffuser flow. AIAA J 1983;19(11):1386–93.

42. Henne P. The case for small supersonic civil aircraft. Reston: AIAA; 2003. Report No.:AIAA-2003-2555.

43. Chpoun A, Passerel D, Li H, et al. Reconsideration of oblique shock wave reflections in steady flows. Part 1. Experimental investigation. J Fluid Mech 1995;301:19–35.

44. Coanda H. Device for deflecting a stream of elastic fluid projected into an elastic fluid. United States patent US 2052869. 1936 Sep 1.

45. Kumada M, Mabuchi I, Oyakawa K. Studies on heat transfer to turbulent jets with adjacent boundaries. Bull Jpn Soc Mech Eng 1972;15(88):1246–56.

46. Ostlund J, Damgaard T, Frey M. Side-load phenomena in highly overexpanded rocket nozzles. J Propul Power 2004;20(4): 695–704.

47. Verma SB, Stark R, Haidn O. Relation between shock unsteadiness and the origin of side-loads in a thrust optimized parabolic rocket nozzle. Aerosp Sci Technol 2006;10(6):474–83.

48. Verma SB, Oskar H. Study on restricted shock separation phenomena in rocket nozzles. Reston: AIAA; 2006. Report No.:AIAA-2006-1431.

49. Lawrence RA. Symmetrical and unsymmetrical flow separation in supersonic nozzles dissertation. Dallas: Southern Methodist University; 1967.

50. Stark R. Beitrag zum Versta¨ ndnis der Stro¨ mungsablo¨ sung in Raketendu¨ sen dissertation. Aachen: RWTH Aachen University; 2010.

51. Stark R, Wagner B. Experimental study of boundary layer separation in truncated ideal contour nozzles. Shock Waves 2009;19(3):185–91.

52. Ge JH, Xu JL, Wang MT, Mo JW. Prediction of flow separation in asymmetric ramp nozzle. Acta Aeronautica et Astronaut Sin 2012;33(8):1394–9 Chinese.

53. Verma SB, Stark R, Haidn O. Reynolds number influence on dualbell transition phenomena. J Propul Power 2013;29(3):602–9.

54. Verma SB, Haidn O. Cold gas testing of thrust-optimized parabolic nozzle in a high-altitude test facility. J Propul Power 2011;27(6):1238–46.

55. Frey M, Hagemann G. Critical assessment of dual-bell nozzles. J Propul Power 1999;15(1):137–43.

Index